Sigma Xi
The Scientific Research Society

Ethics, Values, and the Promise of Science

Forum Proceedings

February 25-26, 1993

Introductory Note: The following papers were prepared for the Sigma Xi Forum in February, 1993. In most cases, the authors submitted a smoothed version of their presentations. In a few instances, the talks were transcribed and revised by the speaker. In addition, the style of referencing in each paper is that used by the author.

ii

All orders must be sent in writing to:
1993 Forum
Sigma Xi, The Scientific Research Society
P.O. Box 13975
Research Triangle Park, NC 27709
Orders for single copies must be accompanied with prepayment of $18.50
plus $2.00 for shipping and handling.

Printed in the United States of America by Edwards Brothers, Inc.

Acknowledgements

Special thanks to the members of the Forum Steering Committee for their expert guidance throughout this entire project.

John Ahearne - Executive Director, Sigma Xi; former Commissioner and Chairman, U.S. Nuclear Regulatory Commission.

J. Michael Bishop - 1989 Nobel Prize-winner in Physiology or Medicine; Professor of Microbiology, Immunology, Biochemistry and Biophysics, University of California, San Francisco.

Sissela Bok - author of *Lying: Moral Choice in Public and Private Life* and *Secrets: On the Ethics of Concealment and Revelation.*

George Bugliarello - President of Sigma Xi (1992-93); President, Polytechnic University.

Rita Colwell - President of Sigma Xi (1991-92); President, Maryland Biotechnology Institute.

Harold Forsen - Senior Vice-President and Manager, Bechtel Technology Group, Bechtel Group, Inc.

Joshua Lederberg - 1958 Nobel Prize-winner in Physiology or Medicine; former President, Rockefeller University.

David Packard - Co-founder of Hewlett-Packard Company.

Frank Press - President, National Academy of Sciences; Institute Professor Emeritus, Massachusetts Institute of Technology.

Chang-Lin Tien - Chancellor, University of California, Berkeley.

Robert White - President, National Academy of Engineering; former Administrator, National Oceanic and Atmospheric Administration.

Sigma Xi extends special thanks to the 1993 Forum financial supporters:

Abbott Laboratories
The Bechtel Foundation
Carolina Power & Light Company
Ciba-Geigy Corporation
Corning Incorporated
The Electric Power Research
    Institute
The General Electric Foundation
Glaxo Inc.
The Johnson Foundation
The Lucille P. Markey Charitable
    Trust

Monsanto Company
Office of Naval Research
The David and Lucile Packard
    Foundation
The Alfred P. Sloan Foundation
Texaco Inc.
U.S. Department of Energy
U.S. Environmental Protection
    Agency
The Weingart Foundation

Contents

Foreword by *George Bugliarello, Sigma Xi President*

The end of this century is witnessing dramatic social and political transformation, of which science and technology are a significant and at times determinant factor.

It was inevitable that in turn science and technology would be profoundly affected by those transformations — that old compacts between science and society be questioned, that the values guiding the scientific and technological enterprise be reexamined and that the inner social dynamics of the enterprise be under pressure to accommodate the new realities.

This volume contains the proceedings of a forum that Sigma Xi, The Scientific Research Society, convened in San Francisco in February 1993. The forum was held in conjunction with the annual meeting of the Society. It is the second in a series of such initiatives that Sigma Xi is undertaking to examine crucial issues at the conjunction of science and the rest of society, that must be addressed if science is to remain the engine of human progress and adventure and to maintain its moral leadership into the next century and beyond.

With over 500 chapters and clubs at universities and research laboratories and in industry, both in the United States and in other countries, and with nearly a hundred thousand members, Sigma Xi is a unique organization that involves science and engineering at the grassroots.

We hope that the topic of this forum will trigger discussions and actions among all the chapters and clubs of the Society and spread from them to their host institutions and beyond. We also hope that this and the topics of the whole series of Sigma Xi fora will be a part of a continuing and expanding dialogue on the role and responsibility of science and engineering in our society.

Introduction

John F. Ahearne
Sigma Xi Executive Director

Sigma Xi in its second century, in addition to its traditional role of honoring research, is focusing on three areas:

- science, math, and engineering education;

- global change and human development; and

- the ethics and values of research.

Recently, a *Washington Post* article, entitled "Louis Pasteur and Questions of Fraud," headlined the charge: "By Today's Standards, the Famed Researcher Committed Scientific Misconduct." The article referred to a talk by Princeton history professor Gerald L. Geison, identified as a leading Pasteur scholar. "...Pasteur's message for contemporary science, Geison argued, was to puncture the 'hopelessly misleading' image of science as 'simply objective and unprejudiced,' a myth that scientists have perpetuated in order to advance their work and attain a 'privileged status.' "

This article is only another indication of why the ethics and values issues are prominent.

Many of the ethical and political debates that capture news headlines have their genesis in the development of science and technology. The frontiers of science and technology provide new possibilities, unimagined in previous generations, but with these come agonizing choices.

At a time when researchers can play a key role in addressing the ethical implications of certain scientific and technological advances, there is a call for science to put its own house in order. "Cases of outright fraud and waste, sloppy research, dubious claims and public bickering have made science an easy target for its critics," declared the August, 1991 cover story in *Time* magazine, "Crisis in the Labs." "Underlying the current furor over funding...," the article went on to say, "are the implicit assumptions that science can no longer be fully trusted to manage its affairs and that society should have a larger voice in its workings...the budget constraints are a part of an even deeper problem afflicting American research: Congress is reflecting an erosion of public confidence in a scientific establishment that not many years ago could seemingly do no wrong. The message from Washington is clear: science will receive no more blank checks and will be held increasingly accountable for both its performance and its behavior."

According to Phillip Griffiths, Director for the Institute of Advanced Study, "Changes taking place in public perceptions and in the Appropriations Committee on Capitol Hill warn of a possibly imminent decay of our university system, the greatest in the world. The debate is a reflection of increased skepticism towards public institutions in general."

The 1993 Sigma Xi Forum planning conference, held in August, 1992, identified many important issues relating to ethics in research:

- Conflicts of interest, or perhaps better described as managing competing interests. These include the more traditional issues of research versus teaching, but also acknowledge the growing concerns about consulting work affecting university time, the dual roles of researchers as industry scientists and academic researchers, and tensions between family life and research life.

- The treatment of postdoctoral students, and mentoring practices, which will determine the attitudes of the next generation of researchers.

- Challenges to the peer review process. As more scientists compete for limited dollars and as research facilities become extremely expensive, the strains on researchers and their students become great. Challenges to the ethics of the peer review process include holding up papers or grant proposals until one's own paper or proposal is ready for submission; related challenges to the tenure selection and promotion process concerning confidentiality of records; and the promises, either made or implied in applying for grants, of what the work will lead to. A senior researcher and administrator said: "We have sold our souls to get funding."

- What responsibilities do individual scientists and the scientific community have with respect to the misuse of science? Scientists are called upon for congressional testimony, to serve on advisory committees to the federal and state governments, and to testify in courts as expert witnesses. What challenges do these activities pose and what responsibilities does the scientific community have?

- The promises of science are great. Are there any legitimate concerns about where science can take society? For example, is dial-a-child, via genetic engineering, both a possibility and a concern which science must address? What ethical issues are raised by the possibility that medical science may one day be able to control the aging process?

- The Environmental Protection Agency, the Food and Drug Administration, the Occupational Safety and Health Administration, and other federal agencies regulate products and work place practices. These regulations are aimed at providing adequate protection against health hazards. Are they adequate, overly constrictive, or too lax? Is this a misuse of science? What responsibility does the scientific community have for the use of resources in these areas? If health is harmed or health effects occur that could have been prevented because of misallocation of resources, is this a question of scientific misconduct if scientists do not speak out or science is misused?

- Howard Schachman and Keith Yamamoto, in a dissent to the recent National Research Council report on scientific misconduct, wrote that "...questionable practices raise issues about the value system and culture of science, and underscore the need for explicit dialogue and education."

- In addition to illuminating the value of science, the issues of misconduct must be addressed. Misconduct must be carefully defined. Procedural guidelines must be developed to ensure both expeditious examination and resolution of charges of misconduct, but also to provide adequate protection both to those who raise such charges and those who are charged. Few whistle blowers go unscathed, but not all whistle blowers are right.

In the year 2000, will we look back and ask why we did not recognize the need to better guide research? Will we see research stifled because of public mistrust? Will we see major opportunities missed because of over regulation or of peer review? Will careers have been ruined because government, media, the public and even scientists could not differentiate between poor work and unethical work? Have the standards of search for truth given way to search for wealth?

The basic issues to be addressed are critical to the success of the research enterprise. Those of us in science believe that the success of the research enterprise is essential for society's future. The scientific community must take a more active role in resolving ethical issues both in society and within the scientific enterprise. The response of researchers to these issues will impact the future of society as profoundly as science and technology itself.

In preparing for this Forum, invaluable aid was provided early by the Sloan Foundation and the Johnson Foundation, which enabled us to have a planning session at the Wingspread Conference Center. Many of the ideas for this Forum came from that session, as did several of the speakers. The overall development of the program came from the Steering Committee, whose members devoted significant time to ensuring a successful meeting. In addition to Sloan and Johnson, the other funders, identified in the acknowledgment, are truly those without which this proceedings volume and the Forum itself would not have been possible.

The papers included in this volume, along with the Conclusions and Recommendations from the breakout groups, provide stimulating ideas for both the research community and the public to address. Sigma Xi intends this Forum be a step in development of a program in ethics and values in research that will be part of the Sigma Xi Center in Research Triangle Park, North Carolina.

Science at the Crossroads

George Bugliarello

1. Today's Predicament

There can be little doubt that science — the scientific and technological enterprise — is at a crossroads (1). Old compacts, spoken and unspoken, between science and the rest of society are questioned by a world that often has seen hopes for social progress deluded in spite of science's magnificent achievements and promise. The need for new compacts and new directions is made imperative by the growing imbalance between those achievements and the ability of society to use them effectively.

When we talk of the promise of science, we must of course be clear as to who does the promising, and to whom. For instance, the Leninist-Stalinist ideology promised a better world through the application of "rational" scientific tenets. That ideology had not been shaped by scientists and engineers but by political theorists and appropriated by totalitarian leaders as an instrument of power. Tragically scientists and engineers had to acquiesce to it in spite of the fact that the Marxist doctrine saw the concept of truth in such self seeking relative terms (Murphy) as to be fundamentally antithetic to science. In the West, during and after World War II and most explicitly after Sputnik, it was primarily the scientists and engineers who promised and convincingly demonstrated the military strength achievable through science. And today science and engineering have become for many an article of faith as the indispensable ingredient of competitiveness and job creation, just as it has for ever greater medical advances.

Each of these promises and hopes had a valid underlying rationale and became embodied, implicitly or explicitly, in some kind of compact. But because each promise has been taken too literally or tied to a specific context, it was inevitable that it would be broken when the context was dishonest, as in the case of science under the Soviet dictatorship or when it changed, as is the case in the United States after the end of the Cold War.

Yet few would disagree that the potential of science and technology is greater than ever. Developments of new scientific ideas and of medical and engineering knowledge are ever more rapid and their societal consequences ever more revolutionary. Suffice it to think of molecular biology or information technology.

Hence, the peculiar nature of what has brought us to today's crossroads and the reason for this Forum: the skepticism engendered by promises, however ill-defined, that have not been realized or seem no longer relevant, and at the same time the frustration of those both inside and outside science, who see the new possibilities offered by science, medicine and engineering but have not been able to overcome the obstacles to transform them into reality. To further complicate the issue, science has become, in the eyes of some, less believable when parts of its vast establishment are self-seeking or individual scientists bring a vested interest to a controversy. An example is the impact of the widespread criticism of self referrals in medicine or of the loss of trust in scientists when a few of them fudge or fake their results or when they contradict each other as expert witnesses.

This is certainly not the first time that science and technology have reached a crossroads in their long path of development that originated when we emerged as a distinct species. Since the beginning of our species, our twin quests to understand and to modify nature have enabled us to transcend some of our biologically inherited or culturally evolved forms of behavior. They have created societies that complement, enhance and at the same time constrain biology and that extend the niches of our survival in our "great drama of fighting against the unknown" (Kranzberg and Pursell). The most compelling reason for examining today's crossroads and for seeking new compacts is the fact that our transcendence of biology through social adaptation abetted by technology is far from successful, as witnessed by today's continuing violence and global environmental deterioration.

The crossroads at which we find ourselves are crossroads of purpose, of method, of organization, of public perception and confidence, and of connection with the rest of society. They are extremely complex because never before have science and technology played so pervasive a role in human society. Neither has there been ever before so urgent a need to address simultaneously so many issues and to make so many choices. These issues are not always present in the mind of all scientists, engineers and physicians and are not always clear in their implications to society as a whole. Let me identify some that appear to me to be among the most important:

- The question of progress, historically and prospectively

- The question of the "end of science" and the related questions of reductionism and of the redefinition of science

- The question of the purpose of science, of the direction of society and of the reconvergence of science and belief

- The question of accountability and, with it, of the support and supervision of science and of a new compact between science and society

- The question of multicultural influences on science, now that science has become a global enterprise

- The question of the specific directions of science and of the restructuring of the scientific enterprise

Before discussing — by necessity very briefly — each of these immense questions, let me underscore what I believe are their two fundamental and inextricably intertwined underpinnings: The ethic responsibility of science and the belief in the further evolution of humankind. Society is doomed if science and technology, with their overarching power, do not have a clear moral sense of their responsibility. But society, so far extended beyond the biological survival capabilities of the individual, is also doomed if it loses confidence in its ability to advance its understanding of nature and in its power to modify nature.

2. The Question of Progress

A belief in progress accompanied scientific discoveries and technical inventions from the Encyclopedists of the XVIII century to the eve of World War I, unshaken even by the carnage of the Napoleonic wars or the American Civil War. The two World Wars and above all the nuclear weapons race shook that belief among many scientists and non-scientists alike. Science and technology came to be seen as having centuplicated the destructive tendencies of our psyche and the negative impact of our species on the planet. That negative sense is reinforced by the ecological catastrophes in Eastern Europe, even if they occurred under regimes that held different views of the meaning of a human life.

Pessimism also has to do with science and engineering having given birth to technologies so pervasive, like autos, television or computers, that they seem to have acquired a power of their own over those who use them and are caught in their pathologies without the possibility of escape. Kurt Vonnegut's autobiographical *Player Piano* or Kobo Abe's works are eloquent visions of people trapped by modern life.

Technologies have given us unprecedented powers, but their undisciplined and indiscriminate use has created the modern existentialist nightmares in which, as in a traffic jam, the victims are at the same time the perpetrators. Diminished as we are when we are caught in these nightmares, we find our lives at risk, economically and even physically, if we do not participate in the technologies that cause them. Such for instance is the fate of the homeless.

It is the revulsion against these historic catastrophes and these nightmares that has brought about more than anything else the loss of a sense of innocence about progress and the promise of science. Particularly disquieting is that loss among scientists. It manifested itself even in objections to the use of the term "the promise of science" in the title of this Forum.

Yet, it is obvious that without science and technology life expectancy today would be much lower — as it was at the turn of the century and as it still is in third and fourth world countries — and life for a large portion of humankind would be much more miserable and unproductive, if possible at all. Thus it is a moot point to blame science and technology for today's ills, particularly since not all societies share — or share to the same degree — our malaise about the future. Most recently, for instance, on visiting China, Stephen Cutliffe observed "little disillusion with the power of science and technology," even though, among some scholars there was "a certain openness to the potential shortcomings of Western Technological development" (Cutcliffe). What is not moot, however, for scientists and engineers is to try to reestablish society's belief in progress. The first step in that direction must be the restoration of confidence among scientists and technologists themselves. That cannot occur without changes in the way science in the broadest sense sees itself and its responsibility to the rest of society.

3. The Question of the End of Science

Crucial to the issue of progress is the *deja vu* notion — restated by some today in *épater le bourgeois* terms — that science may be coming virtually to an end because there is little left of fundamental value to discover (e.g. Elvee).

The reduction of reality to simpler models started with Galileo and Descartes and became the methodological base of modern science. But the notion of the end of science is the last statement of that reductionistic dream that has made periodic reappearances, since the 1600's, becoming also related more recently to the positivism of Comte, Mach and others (Agazzi).

Today's argument must be considered very seriously. It stems from the very important pursuit of a unified theory capable of explaining the origin of mass and nature (Weinberg) and also from the absolute belief by some biologists in the finality of the Darwinian scheme of evolution (e.g. Meyr). Extreme interpretation and extrapolation of these final theories could see science as being left with only "a few loose ends" (Horgan). The importance of the question for science lies not only in the possibility that this would make science revert to a constraining and demoralizing Aristotelian construct, but also that it would keep science focused exclusively on the understanding of nature *as it is*. It can be argued that, even if a final theory were to emerge, the new frontier of science will lie in ending the separation between knowing and modifying nature. This is already happening when we modify nature in order to understand it, as the chemist does now with synthetic molecules, and when we try to understand nature in order to modify it, as the biotechnologist does with molecular biology. (It is tempting to quote what Heidegger said of one science: "Anthropology is the interpretation of man that already knows what man is and hence can never ask who he may be.") The challenge of creating what is not, whether new organisms or new worlds in space, offers science not an end but an endless future.

Few scientists or engineers have debated with sufficient vigor the reductionistic argument of the end of science, although viscerally opposed to it as they see the immense complexity associated with higher levels of organization. That complexity cannot be understood just in terms of the properties of the basic elements of the organization. There is a need for a broad scientific debate to address the issue and again give science and scientists a clearer sense of their future. Otherwise the concept of the end of science, misinterpreted or put forth in hubris, can only deflect from science the brightest minds on which it relies for its advance. Denial of reductionism does not mean, however, that the search for unity in science should be abandoned among the great varieties and directions of scientific pursuits. As Francis Bacon underscored quoting an old ecclesiastical statement: "in veste varietas sit, scissura non sit" — let there be variety, not schism.

4. The Question of the Purpose of Science and the Direction of Society

Science and technology have made possible today's society — a society far different from that envisioned by the Encyclopedists, by Rousseau or Hobbes, or even by our own Constitution. By enhancing beyond measure the material aspects of our society, science has profoundly affected the relation between the spiritual and the material that is fundamental to our humanity. Twentieth century culture is polarized between an exasperated quest for rationality and an irrepressible emotionalism. The first is expected to govern our actions and our technological-based society. The latter manifests itself in wars, in a search for fulfillment through religion or cults and in an aggressive, popular and often self-indulgent post-modernism. The very visible embodiment of the polarization in today's architecture serves to remind us ourselves of how much closer instead to art was modern science at its dawn (e.g. Siraisi). Today science and technology have become divorced from art even though they may search for beauty in the structures of nature and in utilitarian artifacts (e.g. Hoffman and Torrence). I believe that a reconvergence of science and art would go a long way toward the reintegration of science and culture.

But above all, in their relentless pursuits, science and technology have weakened the aspiration in our lives to something that transcends the material and that provided a moral coherence to previous societies, as in the Middle Ages (e.g. Cantor).

So, having defeated the grotesque scientific pieties of Marxism, and at the same time questioning the concept of progress, pragmatically as well as on religious or philosophical grounds, our society is now moving over uncharted territory, torn between the two polarities. Ironically, the emotional reaction to science and technology is also a consequence of the immense production capacities that scientific technology makes available to satisfy our primitive instincts. With consumption the bellwether of well being, society has become unable to envision and rationalize its own limits and to discipline the acquisitive urges of its members. How many pairs of shoes are enough? How many pairs of jeans? How many cars? How many appliances? Indeed, how many of us? And, is there any purpose to the relentless growth of our physical stature? Most importantly, how does one reconcile our unchecked urge to continue with the poverty of a very large portion of mankind?

Thus, ours is a society truly at the crossroads — of which science's own crossroads is both the cause and the consequence. If some 500,000 years ago the utilization of fire became a key divide in human evolution, today's science and technology's ability to alter our environment and our biology to a degree unimagined even in the middle of the century, are an even greater divide.

At this junction the burden is squarely on scientists and engineers. Can we help provide intellectual and moral leadership to a society that cannot continue without great peril to be torn apart between today's polarities? Science — the scientific and technological enterprise — has caused or abetted that polarity. It must now decide whether it has the capacity to transcend its self-imposed boundaries and help guide not only with rationality but also with passion and compassion a human society that in recent years, thanks to science and technology, has acquired a common history and a common future. This is an unprecedented challenge for a science that has derived great advantages from its interactions with government and industry, but has been immensely reluctant to assume deeper responsibilities and enter the arena of politics for which it holds such revolutionary implications (e.g. Bondi). Within science itself, the very conservatism of the scientific collective mind may often discourage expression and discussion of ideas yet only imperfectly formulated. That discussion, however, is essential today as it was at the time of Copernicus and must be fostered if science is to transcend its traditional limits. Hence, again, the importance of this Forum.

5. The Question of Science, Religion and Philosophy

Politics is only one aspect of today's dilemma for science. The deeper challenge is whether a way can be found to overcome the obstacles that at different times and for different reasons have created a seemingly impassable chasm between science and religion as well as the moral aspects of philosophy. From its early times humankind has believed in an afterlife. By destroying that belief without providing some equivalent comfort, science will not be able to point society toward the future unless it becomes more emotionally satisfying and better adept at communicating the passion, the joy and the comfort of its pursuits.

Also, from time immemorial humans, both individually and in entire societies, have striven for greatness. That quest has been neither universal nor explicit. But for the first time there is through science the opportunity to offer, not to selected groups but to humankind as a whole, a new vision of greatness based on a new vision of the future of man. We scientists and engineers have the potential, if we are willing to reach boldly beyond the comforting fastness of science's traditional domain, to prevent today's squandering of human talent in the pursuit of a globally unsustainable consumerism. We have the potential to stimulate a new vision and a new philosophy of the law to better deal with the modifications of our biological nature, with our responsibilities to the rest of the universe as well as with the role of the human-made in our lives. And we have the potential, if united in our purpose, to move both government and religions to address the questions of the limits to population growth and of global responses to war, poverty, disease and ignorance. In this can lie the ultimate greatness of science and of our species.

It would be extreme hubris, however, for scientists, engineers and physicians to believe that we can solve these problems by ourselves. But is it not moral cowardice for us not to use our resources of knowledge and know-how to propose new paradigms of the relation of the biological, the social and the machine that will stimulate society to rethink these crucial issues (2)? And is it not also possible for us to go one step beyond, by becoming passionately involved in the required decisions and in their execution? If we choose this path, the starting point must be the conviction that it is legitimate and urgent for science to search for a new morality and for a new alliance of science and belief. That search is too important to be left to the philosophers and the theologians alone.

After 200,000 generations of humans as a distinct species, the new potentials of science project the possibility of genetic and reproductive interventions outside the barriers of traditional biology. In so doing, they strike at the core of many religious beliefs and demand a fundamental rethinking of the relation of science to religion — religion as a powerful formal social entity but also as the expression of an innate human sense of wonderment at creation, of hope beyond death and of one's responsibility to one's self, to other humans as well as to nature. Science can acquire a religious connotation in the measure that, on the basis of its understanding of nature, it seeks new ethical rules for humans and human society. The starting point in rethinking the relation of science to religion may be the recognition that emotions and belief are essential tools in building societies, civilizations and cultures, just as science and technology are. Bertrand Russell's position that science cannot decide on value problems is simply too limiting. Although it is true that such problems are outside the realm of truth and error, as he puts it, they can benefit from the involvement of scientists and technologists who are sensitive to the importance of emotion and belief and capable of accepting them in applying the results of their investigations.

The outcome to be avoided is a renewal of the "warfare of science with theology," to use White's famous expression. Certainly Einstein did not see science and religion as "irreconcilable antagonists" — yet that warfare has been much too real, making it so difficult even today in many parts of the world to address reproductive issues and the connected ecological ones. Thus, it is to be hoped that it may be possible after millennia to see again convergence of science and belief. This does not mean closing the circle for a science that was originally the instrument and part and parcel of religion, but it means endeavoring to bring science and religion on an openly synergistic path motivated by the common desire to enhance and elevate the human condition.

If this is the path to be chosen, it puts an enormous responsibility on science and the scientists, as it does on religion and religiosity, not to view science — as Pascal ultimately did — as antithetic to the concept of spirituality. The Soviet state was the most recent and dramatic example of the dangers of a science and technology devoid of moral compass and operating as instruments of an exclusively materialistic doctrine. The lesson is not the collapse of that state, as it is entirely conceivable that, with different leadership and better organization, it could have prevailed, just as the might of the Mongols, with their primitive but most effective military technology, prevailed in the XIII and XIV centuries over much of Asia and Europe. The lesson, rather, is the destruction of civilized forms of human existence engendered by a science and technology unwilling or unable to take a strong moral stand. In our society, indifference to the plight of the cities, the poor and the sick, or to the trampling of privacy are a clear sign of danger and a moral warning to our science and technology not to sit idly by.

We must note of how little help, in these issues, are the specific philosophies of science and of engineering. The crucial philosophical problem we need to address has to do with our purpose in endeavoring to understand and modify nature. Today the philosophy of science is focused far more on the concepts, paradigms, theories and methodologies of science, than on that purpose. The philosophy of engineering, much more recent in its emergence, as of late has been drawn almost inevitably, given the urgency of the need, to the issue of the public responsibility of the engineer (Mitcham). But the central question for engineering is that of the nature and purpose of the artifact; in spite of some groundbreaking research work (e.g. Durbin), it requires much more attention. The growing field of bioethics, in considering the moral health care dilemmas of cost and access, is the one that comes perhaps closer to addressing the question of the modification of nature, albeit in the specialized context of medicine (3). Clearly any philosophy of those modifications, whether one deals with engineering or medicine, must address the issue of permissible future directions of the modifications. To search for what is humanly desirable demands a deep dialogue of science with belief and philosophy. The dialogue is urgent, but will not occur unless it is made an integral part of the education of scientists, engineers and physicians.

6. Compacts, Covenants and the Accountability of Science

The concept of a compact or, more formally, a covenant is of course not new in human history. The ten commandments, the Magna Charta, Rousseau's social contact, Hobbes, and more generally of the compact theory of the XVIII century and the ensuing concept of natural rights embodied in our Constitution (e.g. Windelband) are all examples. However, the belief in a compact in force between science — broadly defined — and the rest of society is a recent one. It has its origin in the U.S. of World War II when science was called to help win the war. Science of course had been called to help the war effort in other nations and other conflicts as well. But only in the U.S. did there emerged a clear statement of the unspoken pact between science and the rest of society, as articulated by Vannevar Bush. The pact was reinforced in the post war years by the activities of the Committee on Science and Technology established by the House of Representatives in 1959 under the stimulus of Sputnik ("Toward the Endless Frontier"). That unspoken pact has come to be viewed by many scientists in the U.S. as a solemn commitment by society to support science, even if never formally stated and even if its original purpose was military, while today our society faces many other challenges.

Because of this long lasting, largely unexamined and ultimately naive belief in the self evidence of an obsolete compact that is now being invoked under very different circumstances, we scientists and engineers are shocked when the compact is questioned by the rest of society, and are unprepared to propose a new one. Yet a new compact is needed. Our society cannot function and prosper without the support of a strong science base. Science in turn cannot expect to thrive and help guide society if it does not consciously and explicitly address the question of its relation to the rest of society and of its accountability.

The challenge in formulating a new compact is complex. In the first place, science must accept as legitimate the concern of society as to how science — with its immense power — keeps its own house. In turn society, in committing itself to supporting science, should not deprive science of the freedom so essential to its health and success. The genius of a compact is in preserving this delicate balance — a balance that, to paraphrase Dante's definition of the law, if preserved preserves human society, but if destroyed destroys it.

Secondly, it is evident today that a new compact must find a way to respond not only to the new needs of defense but also to the harder to define ones of commercial competitiveness and social progress. So much is being discussed about these needs, as not to warrant further elaboration here.

In the third place, a new compact must deal with the supervision and financing of science and with conflicts of interest. Inevitably, as science comes to absorb significant portions of a nation's Gross National Product, there is bound to be an increasing demand by government and the public for strict accountability of scientific expenses. Furthermore, as research is viewed more and more as a key ingredient of competitiveness, its direction and its potential conflicts of interest become major issues. The situation is quite different from the old compact when science in the U.S. had substantial freedom to pursue its own goals and to police itself in what many scientists believe was a golden age. The question today is whether the scientific enterprise is capable of taking a hard look at itself and policing itself so as to reduce the need for outside intervention.

Fourthly, the not always politically correct question should be addressed as to how science will also fulfill its responsibility to itself, as science for science's sake — science, as Francis Bacon put it, for ornament and delight, rather than exclusively for utilitarian purposes.

Lastly and most importantly, if a new compact is to be formulated, we must decide whether it should be based on an even broader sense of reciprocal responsibilities of science and society than outlined in the previous points. The first of those responsibilities is to restore a sense of hope and progress. I already suggested the importance of a new relation of science to belief. Should a new compact — indeed a new covenant — be governed also by a new set of fundamental concepts involving our relation as individuals and society to the artifacts we have created, and to the rest of the nature — the environment on earth and beyond? If so, should not the covenant, in addition to providing a practical guide for the relationship among these entities, also be inspired by a new vision of how that relationship can evolve to enhance our essential humanity?

7. The Question of Multiculturalism

If science endeavors to establish a new ethics for itself and a new compact with the rest of society, one issue that looms large but has received scarce attention is multiculturalism. The second half of the century, after World War II, has seen an embracing of modern science by wider and wider groups of the world population. Science has become virtually universal. It is practiced, albeit with different intensity and success, in most parts of the world, and looked upon as an instrument of progress and disenfranchisement from poverty as well as, much too often still, as an instrument of military power.

Scientists all over the world are engaged in what for all practical purposes is one science, Western science, even if created by the historic contributions of several cultures. Yet the cultural beliefs of the scientists — ethical, social and religious — are obviously far from uniform. Thus an extremely powerful and universal instrument, science, is not wielded today with a commonality of purposes, ethical views or sociological practices. Under these conditions, the emergence globally of a new role and a new responsibility for science becomes that much more difficult and the accountability of science to society that much more complicated. The problem is urgent and must be addressed. The issue of multiculturalism applies also inside the U.S., where a substantial portion of academic researchers come from many different cultures. If successfully resolved, the very challenge that U.S. science faces in this regard can provide a unique laboratory and model for the global community. In that model multiculturalism, rather than being submerged or ignored, could become a new powerful instrument of science.

8. The Question of the Specific Direction of Science, of its Organization and of the End of the Separation Between Knowing and Modifying Nature

An ever present question in science is the direction in which it will move at any given time, under the momentum of its own discoveries and achievements as well as under the influence of exogenous factors. The question becomes particularly critical today, in a period of great expansion of knowledge and great investment in science. It affects most immediately not only the scientific and technological community, but also the life of every citizen, whether one considers the issue of AIDS, of research into "orphan" diseases, of energy policies and nuclear power or of science education.

Of all the many possible directions in which science can move, those favored by investment of resources and by the attention of the scientists are obviously more likely to advance. Given the limits to the resources available, this places a heavy burden on the scientific community to intervene in the decisions, pondering their implications, not only in the narrower context of the advancement of science and technology, but in the broader one of the interest of society as a whole.

The basic choices for science today — such as unmanned space exploration versus space stations, "big science" versus broad base support of individual investigators, biology and medicine versus applied biotechnology, life sciences versus the physical sciences and engineering — are too well known to bear repeating here, even if some issues such as the support of mature sciences or of far out independent researchers warrant more attention (4). Rather, what needs to be underscored is the issue of whether the traditional organization and institutions of science should be modified or enhanced. For instance, would a reorganization along more agile lines facilitate interdisciplinarity as well as disciplinary advances? Also, would a rearrangement of institutional and disciplinary boundaries facilitate a closer relation of science with the rest of society as well as the integration of the knowing and the modification of nature?

9. The Path Ahead

My purpose in this paper has been to focus, by necessity in a very brief and subjective way, on some of the enormously complex issues that have brought science to today's crossroads — and to underscore some of their implications. If we do not intend to continue with business as usual, we must make a deliberate attempt to place science, engineering and medicine on a higher moral plane — higher not because of authority of knowledge, but because of the association of that authority with a passionate commitment to make the best use of that knowledge.

What is the best use implies of course a value judgment. The most fundamental question at today's crossroads is whether science, *qua science*, should participate in those value judgments. Clearly a value judgment is not a scientific statement. But society's investment in science, and the scientific — technological enterprise's impact on society, are the results of value judgments. We scientists, and the scientific community in its ensemble, cannot disavow the fact that our knowledge and know-how are an indispensable base for the kind of difficult value judgments that our society will be called to make with increasing urgency in deciding the directions of its future. For example, should science have something to say as to whether there be limits to material growth, population growth and consumption or to the prolongation of life at all costs, as to what should be the purpose of space exploration and travel, as to the issue of homosexuality, as to the most appropriate role of automation and hence of work in our society, or as to the allowable direction of biotechnological interventions on our own genes and on those of other species? (Consider for instance the current controversy about the use of tomatoes genetically engineered to delay spoilage).

At this moment, by default, many of these judgments are based on irrational fears, on short-sighted self-interest, on political expedience, on inflexible religious dogma or on economic paradigms that, even if greatly refined and multidimensional, are still the progeny of Henry Adams' *homo oeconomicus*.

The disasters that continue to beset our globe, from the Somali famine to the burning of Los Angeles to ethnic cleansing in Europe, will be repeated and amplified unless the value judgments we make about the directions of our future are guided by a much stronger infusion of scientific knowledge. But for that to happen, sheer scientific knowledge, no matter how penetrating and capable of predicting and modeling future trends, no matter how powerfully complemented by the technological ability to modify nature, does not suffice. Why would we have today so many faith healers, so many tales of the supernatural in books and movies, so many parodies of the scientist and the engineer — the scientist usually mad and the engineer a nerd — if scientific and engineering knowledge were felt to be comforting, believable and emotionally satisfying, a knowledge to which one could both rationally and in faith entrust one's future and with which one could entertain a rich human dialogue?

Unfortunately today we scientists and engineers tend to avoid that dialogue. We talk to ourselves. We retrench behind the intimidating jargon of our trade and the professed avoidance of value judgments, and we eschew public life and the public pulpit. In the U.S., with less than a handful of scientists and engineers in Congress, the direction of the country is left to lawyers and other callings that are removed from any direct experience in understanding or modifying nature. Scientists and engineers confine themselves to cautious advice to the federal government through the National Academies and other organisms, and are satisfied with subordinate positions of power in government. Even weaker is their position in state and local governments, where so many day-to-day decisions affecting our lives are being made.

At today's crossroads it is urgent to decide whether science and technology should endeavor to pursue a new path that will give them a new moral authority and a new role in helping guide society toward a better future. The responsibility for that decision must fall in the first place on American science, as the largest and most influential single grouping of active scientists and engineers in the world — a grouping unhampered by state ideology and by religious dogma and built on the acceptance and utilization of scientists from all over the world.

Notes

Note 1

For brevity, when warranted, I shall use the term science to encompass both the discovery of nature and its modification by rational means, that is by technology — engineering, medicine, etc. Further, although science, scientific method and scientists are not synonymous terms, I may in a context when the distinction is not important use one for all three. The term "science" itself is one of multiple meanings — an activity, a body of knowledge as well as the complex of individuals and organizations that do science. In their endeavors to understand and modify nature, science and engineering nature differentiate themselves from other efforts to understand and modify nature by the method they use — the scientific method.

Note 2

For the sake of brevity, I have used elsewhere the term "biosoma" to describe the entity formed by the indissoluble combination of biological organism, society and machines, i.e. artifacts (e.g. Bugliarello). The interactions among these three entities, and between them and the environment, shape our lives and, as they evolve, determine our future.

Note 3

Medicine in its endeavor to modify a natural phenomenon — disease — is akin to engineering. They both are based on a scientific understanding of nature but committed to modify aspects of nature.

Note 4

It may be argued that mature disciplines in a new context can have a resurgence, as is the case of pharmaceutical botany or of railroad engineering with the TGV's (trains grande vitesse) and the prospects of magnetic levitation. Also it may be argued that it is not clear whether the era of the far out inventor not supported by a social infrastructure of agencies, foundations, study groups or academic department, as was initially the case of Marconi or Goddard, is necessarily over, and whether we need to pay more attention to that precious human venture capital.

References

Agazzi, Evandro, "The Problems of Reductionism in Science," Episteme, Vol 18, Kluver, Dordrecht, 1991

Bacon, Francis, "The Essay or Counsels, Civil and Moral," Peter Pauper Press, Mt. Vernon, New York

Bondi, Hermann, "Bridging the Gulf," *Technology in Society*, Vol. 4, pp. 9-14, 1992

Bugliarello, George, "Technology and the Environment," in "Changing the Global Environment," Botkin, Caswell, Estes and Orio, Editors, Academic Press, New York 1989

Bush, Vannevar, "Science, the Endless Frontier," A Report to the President. U.S. Government Printing Office, Washington, D.C. July 1945

Cantor, Norman, "Inventing the Middle Ages," Morrow, New York, 1991

Cutcliffe, Stephen "Of Auto Horns, Steam Locomotives and Hydroelectric Dams." *STS Today*, University Park, Pennsylvania, Nov. 1992

Durbin, Paul T., "Research in Philosophy and Technology," 12 volumes, Jai Press, Greenwich, Connecticut 1978-1992

Einstein, Albert, "Ideas and Opinions," Bonanza Books, New York, 1954

Elvee, Richard Q. (Editor) "End of Science? Attack or Defend," University Press of America, Lanham, Maryland, 1992

Heidegger, Martin, "The Question Concerning Technology and Other Essays," William Lovitt (Translator), Harper Torchbooks, New York, 1977

Hoffman, Roald and Vivian Torrence, "Chemistry Imagined," Smithsonian Institution Press, Washington, D.C., 1993

Horgan, John, "The New Challenges," *Scientific American*, Vol 267, No. 6, pp. 16-22, Dec. 1992

Kranzberg, Melvin and Carroll W. Pursell, Jr. (Editors), "Technology in Western Civilization," Vol 1, Oxford University Press, New York, 1967

Mayr, Ernst, "One Long Argument: Charles Darwin and the Genesis of Modern Evolutionary Thought," Harvard University Press, 1991

Mitcham, Carl, "Ethics in Engineering Research," Chapter 8 in K.S. Shrader-Frechette, "Research Ethics" (in the press).

Murphy, Kenneth, "Retreat from Finland Station," Free Press, New York, 1992

Russell, Bertrand, "Religion and Science," Oxford University Press, 1961

Siraisi, Nancy G., "Arts and Sciences at Padua: the Studium at Padua before 1350, "Pontifical Institute of Medieval Studies, Toronto, Canada, 1973

"Toward the Endless Frontier - History of the Committee on Science and Technology, 1959-79," U.S. House of Representatives, U.S. Government Printing Office, Washington, D.C., 1980

Vonnegut, Kurt Jr., "Player Piano," Delacorte, New York, 1952

Weinberg, Steven, "Dreams of a Final Theory," Pantheon, New York, 1993

Windelband, Wilhelm, "A History of Philosophy," 2 Volumes, Harper, New York, 1958 (reprint of 1901 translation by James W. Tufts)

White, Andrew D., "A History of the Warfare of Science with Theology in Christendom," 2 volumes, Dover, NY, 1960

The following three presentations are from a panel with the topic: "Ethical Research: Principles and Practices." Each of the presenters is a Nobel Laureate.

Ethical Research: Principles and Practices

Yuan T. Lee

I was born in Taiwan in 1936 and grew up during the Second World War. Toward the end of the war, in the midst of bombing by the Allies in Taiwan daily, we wondered whether we would survive another day. By the time the war was ended, I was in the second grade. As a student, I did very well in science, but I enjoyed playing baseball better. The very reason I became a scientist later on was because in the tenth grade, because of illness, I had to stay home from school and had a lot of time to think about the meaning of life. I always asked the question, what did I want to get from my life. It was at that time, I read the book *The Autobiography of Madame Curie*, and that book inspired and influenced me a great deal. I decided I wanted to be a chemist.

There are two things in the life of Madame Curie which really moved and inspired me. The first thing is the excitement of scientific research. Her dedication, her hard work, the discovery, the excitement. I suddenly realized the scientist could have a beautiful life. There is another thing which is more important to me. It was the idealistic view of Madame Curie. She was a very idealistic person. When someone asked her, "Why didn't you patent all of your discoveries, like the discovery of radium/polonium? If you were to have done so, you could be very wealthy like Edison." Also, during the First World War, she was driving an ambulance, helping the wounded soldiers. In her idealistic view, that knowledge belonged to mankind, she should not use what she has discovered through her scientific research for her own gain.

These two things, the excitement of scientific discovery and the idealistic way of giving of life, were truly what motivated me to want to follow in her footsteps. Of course, in the modern society we also have to recognize two other things. In a modern society, scientific research is really a human activity. Actually, it is a very important part of the cultural activity of human society. What I try to say is the following: society is funding the scientific research. Many of us working together try to accumulate the knowledge, in the sense that the accumulation of knowledge is really the most important part of the scientific research. Of course, after knowledge is accumulated, some day — some time, someone will summarize all of those discoveries in the way of a breakthrough. But as to who accomplishes this breakthrough, it is not that important. The accumulation of scientific knowledge is the most important thing. A breakthrough is bound to happen, somebody will be able to do it. The other thing is this. As scientists, we try to accumulate the knowledge and try to accomplish something — discover something new, find some gold in nature. Another important aspect: although the scientific discovery will eventually become very, very useful in the future, our contribution to the society, when society provides us with funding to let us do scientific research, our contribution really in part is in the training of the well educated young people. In a sense, we find many new things that are very important, but at the same time we train some young scientists, who understand science, to be able to be of service to society. This is just as important. If I were to say that those four points you have to keep in mind to be a scientist, certainly you will ask the question, if you derive the excitement for scientific research mainly because you enjoy the excitement of discovery and you like to work together with bright young scientists, then the pursuit of excellence should be really the search of choice, not the fame or the wealth.

So certainly, I come up with this kind of question: in the modern society, does it make sense to honor a scientist or give prizes, give recognitions, Nobel prizes — does it make sense? The answer to this is kind of obsolete. I remember about ten years ago, when one of my former colleagues, the late Professor Hildebrandt, was celebrating his hundredth birthday. The American Chemical Society gave him the Hildebrandt award. He was the first person to receive this Hildebrandt award. A few people asked him, "Joe, we all understand the motivation which makes us work so hard is really the excitement of scientific discovery when joined together with our young colleagues, why is it necessary to keep on giving awards, recognition?" So I turned to him and said, "Joe, you have lived for a hundred years. During your lifetime you have received so many awards. Does this still excite you, or are you really happy about receiving another award?" He said: "You are still very young (at that time I was young — not any more); you are quite right. Not too many of us work so hard just to get recognition. On the other hand," he said, "not too many people out there understand what we are doing, so it is our responsibility to tell them, tell the society what we have been doing. Giving honor, giving recognition to scientists, is not important to those individuals at all, rather, the one who receives the award or the recognition will have the responsibility to tell the world what he is doing. To tell the population what kind of activities he has been engaging in."

I am a physical chemist. I am an experimental scientist and what I have been trying to do is to try to understand how chemical reactions take place when molecules collide. Of course, my part of the research is so-called closer to the exact science, doing some precise things. When I try to do research with my students, I always try to obey the following rules:

- The first thing is this, if I pick up a problem I always ask the question could that problem be solved by somebody else at the present time — if you exchange some ideas, could somebody solve the problem? In three months, six months, a year or two years? If the answer is yes, then why do you do it? You should discuss with some of the people and somebody should do it. It means that if you want to pick up a problem, to solve it, you really would like to solve a problem that nobody else can do — you are the one who is equipped to do it or you are trained well enough to do it or you have some idea to do it. So you have to have a long view that we are solving the problem all together and you should do something that nobody else can do. This is one thing that I always want to do. Because if I do this, then I will realize what happens on the very frontier of scientific programs. Your student has to be very well trained. We can not use them as a play of hand, they are the ones who will solve the real problems. So the training and the carrying out of research will not have any conflict.

- The second thing is that when we work on the experimental research with students, I'll try to inspire them a bit by setting a good example of a dedicated, hard-working scientist who enjoys scientific discovery and hard work and enjoyment of life.

- The other thing I want to do is this. If I pick up a problem and want to work with students, I always ask the question whether I would be willing to dedicate myself to do it once I set my mind to do it. If I have a relatively large group, twenty people working with me in the laboratory, of course I cannot participate in every project in every detail. But I want to make sure, when we see the problem, that it is a project that I myself would be willing to do from beginning to end.

- The fourth thing I want to tell my students is that knowledge is accumulated by mankind. It is not complete, it is far from perfect. It means that they don't depend on me. I will be able to give general guidance, but they are the ones who really cut through the frontiers and make the new discoveries.

Those are the kind of guidelines I try to follow when I carry out scientific research. Of course, you might say that the kind of thing that you are doing might not be very useful, so you don't have to worry about the wealth in that kind of thing. That might be true but it might not be completely true. What we are doing certainly will have some implications sooner or later. But I do think there is one very, very important question scientists always hesitate to ask, the kind of question one has to ask all the time. I did mention at the beginning that

Madame Curie had a very idealistic view and said that knowledge should belong to all humankind. I think many scientists sitting here would agree with this. As soon as we discover something, we submit a paper to a scientific journal publisher, to let the world know what we have done. We enjoy discussing together, solving the problem together. If we look at the history of mankind, although we are quite advanced, but as we inch toward the end of the twentieth century, we are really not advanced enough. What I am trying to say is this. Although we believe scientific knowledge belongs to mankind, somebody translates the scientific knowledge into scientific technology or social productivity, then that would belong to a certain group of people, or somebody, or some organizations. So there is a certain conflict between science and technology. Science belongs to mankind, yet when you translate scientific knowledge into technology, then it would not belong to all of mankind. Of course, why raise the question? People would tell me that you are raising the question of what is good to propel the advancement of society. I think in the future, maybe fifty years or even one hundred years from now, everybody will work together in technological developments and scientific discoveries. We will be in agreement that knowledge belongs to mankind, and advances of technology and social productivity also belong to mankind, and maybe all mankind will work together and the conflict between science and society will be lessened a great deal.

Rosalyn S. Yalow

The topic of this short symposium "Ethical Research: Principles and Practices" has so many facets that simply listing all of them might take up much of the time allotted to me. Those of us in biomedical investigation have what may be considered a unique problem, that is, the use of living beings as experimental subjects. A half-century ago studies employing people did not require review or even that the people involved be informed of possible hazards associated with the study. For instance a recent VA-funded study conducted by the Institute of Medicine reminds us that in 1941, the U.S. began an intensive research program on poison gases because it was expected that such gases would be used in WWII. The White House organized research programs to test a variety of protective measures. Thousands of military personnel were used as test subjects. More than 4,000 experienced very severe exposures in tests conducted under field conditions or in gas chambers. Overall, the levels of exposure by the human subjects may have been much higher than previously believed, perhaps as high as battlefield exposures, and the people involved were not informed as to the possible hazards or asked about their willingness to participate. So-called research programs such as this cannot happen now in our country. Studies involving people are reviewed by human studies committees. The participants must be informed about the nature of the study, its problems and potential harm and must sign an appropriate statement of informed consent. A recent criticism of current cooperative studies is that women and minorities may not be studied as well as white males, but that is an ethical issue I will not consider now. However it should be noted that an ethical reason for not using women of child-bearing age in drug studies relates to potential harm to a fetus if the woman were pregnant.

Another area of concern is the attack by so-called animal rights groups on the use of animals in the biomedical setting. On occasion these attacks have been so vicious that protection of the people and facilities involved in the work is required. I cannot help but wonder if those who would deny the use of animals in biomedical investigation would deny themselves or their loved ones the benefits to their health and well-being that have resulted from investigations employing animals. Over 90 percent of all animals used in research are mice, rats and other rodents. For the most part we kill mice that invade our homes or rats in our parks without concern for their rights. The most important use for large animals such as dogs and pigs in medical institutions are in surgical training — would you want a surgeon to learn to practice new procedures on you or your loved ones or on a dog or a pig? Over the years there have been a number of Federal laws dealing with care of animals and each institution using animals is required to have an Animal Care and Use committee to assure that all animals are treated in a humane fashion. Hospitals and research institutions must be certified by an independent group, the American Association for the Accreditation of Laboratory Animal Care (AAALAC) to assure adherence to rigorous standards of animal care.

To summarize biomedical investigation using living animals — including people — is well or perhaps even overregulated to assure that ethical problems that were not uncommon 50 years ago do not occur now.

Next I wish to discuss some ethical problems associated with the peer review system for funding research. There is general agreement that accountability must accompany the expenditure of large sums obtained from the government or other sources. The National Institutes of Health, the principal funding agency in the United States for biomedical research, requires investigators to prepare a proposal describing in great detail their research plans for a three- to five-year period. There are many deficiencies in the system — including its inherent dishonesty. For instance, few investigators whose contributions are highly original and imaginative can spell out, as presumably is required in such grant requests, detailed plans for so extended a period. Generally in investigation new studies evolve on the basis of previous results. Furthermore, is it realistic to expect an investigator to reveal to his peers a highly novel original idea that is planned for future work without considering that his priority might thereby be dissipated? For instance, when Hahn and Strassman's observations and the subsequent theoretical considerations on fission were revealed to the scientific community, within days confirmatory experiments were performed in laboratories around the world. Even assuming complete conscious honesty on the part of those with access to presumably confidential information contained in grant proposals, it is impossible to assure that knowledge of a very original idea or finding disclosed to a restricted few does not confer on them an unfair advantage.

Behind the scenes, there are those who say that the system does not really work the way it appears. They claim that everyone knows that what an investigator describes in his proposal is not his plans for future investigation but is rather work in progress that may well be completed even before the grant is awarded. Alternatively the investigator may devote a portion of his time and funding to the proposal and surreptitiously bootleg the rest of his effort to the preparatory work for the next grant proposal. Such a system is fraught with disastrous consequences. If the investigator honestly adheres to the grant proposal, he is not free to take advantage of new ideas developing out of the studies; if the investigator resorts to the various subterfuges just described it raises the question as to whether honest science can be expected to result from such devious tactics.

A system based on a retrospective review should be considered. Funding at a constant level could be renewable, perhaps at three-year intervals, subject to submission before termination of the grant period of a satisfactory progress report to demonstrate that effective use has been made of the funds. If it is evident that productivity has significantly decreased, funding could be decreased or discontinued. If significantly increased funding is requested, justification should be on the basis of a significant increase in the scope of the project, which might then require additional review.

It is important that the entire research effort of a scientist or group of scientists be examined at one time. The ruse of submitting a multiplicity of proposals to different sources seems to make each part of an interrelated investigation look less expensive, leads to increased paperwork, makes it impossible to evaluate properly the integrated effort and does raise questions as to the ethical problems in the search for funding.

Let me turn now to the problem of fraud. Obviously, fraud in science is inconsistent with excellence or even with mediocrity — it simply is not science. I believe that those who commit fraud, defined as the wholesale invention of data, are rare in the scientific community and probably have some degree of psychopathology. Those like Darsee in the laboratory of Eugene Braunwald or Summerlin in the laboratory of Robert Good were working in important areas of investigation and should have appreciated that the fraudulence of their data would eventually be detected. Therefore, if they had considered the situation rationally, it should have been evident to them that such deviant behavior could not lead to a brilliant future in research. I am less concerned with the aberrations of the junior perpetrators than with the attitudes and behavior of the senior scientists in whose laboratories these events occurred.

Outright fraud is generally eventually detected. But there is concern about the multiple forms of dishonesty that might be quite subtle and more common in laboratories that are not carefully supervised. The integrity of the research process depends ultimately on self-regulation. All of us who engage in the search for knowledge must accept full responsibility for the supervision and training of the young investigators in our laboratories. We must set an example of intellectual honesty. We must neither exploit them nor take credit for their work. We must evaluate them wisely and encourage those possessing excellence, imagination and originality and discourage the pedants among them. Incidents of dishonesty in science will continue to occur until senior scientists understand that if there is unethical behavior in their laboratories, it is they who are personally responsible, they who share the guilt, and they who cannot evade the onus. Perhaps every research director's desk should carry President Truman's motto, "The buck stops here."

The last issue I will simply present to you for your consideration but will not have time to discuss. In 1953 Irving Langmuir, a 1932 Nobel Prize winner in Chemistry, presented a colloquium on Pathologic Science which was republished recently in the October 1989 issue of *Physics Today*. Pathologic Science in physical science is characterized by investigators who are not dishonest but who report selective results, often over an extended period of time. The effects described are barely detectable but are claimed to be reproducible and of considerable significance. Langmuir wrote about N-rays, first reported in 1903 by Blondlot, a member of the French Academy of Sciences. N-rays were described as having characteristics between X-rays and α-rays which were already known to be important. After a visit to the laboratory by R.W. Wood, it was demonstrated that N-rays were no more than a product of Blondlot's imagination. About 1923 Gurwitsh of Moscow described mitogenetic rays that were presumed to be given off by growing plants and other living things. These rays could pass through quartz but not glass. It took more than a decade for the hundreds of reports describing the non-existing mitogenetic rays to end. But the most amazing story of all was the Allison effect which was first reported in 1927. There were hundreds of papers published in respectable journals such as the *Physical Review*, *Journal of the American Chemical Society*, etc. describing different new elements identified with the use of the Allison effect as well as many other studies by very respectable scientists. The new elements, including alabamine and virginium, were considered so important that they were reported in *Discoveries of the Year*. A decade later the Allison effect could not be reproduced and no more papers concerning it were published. The so-called Allison effect is probably unknown to the post-World-War-II generation of scientists.

At present there are fields attracting considerable attention that I believe will sometime in the future be identified as Pathologic Science. These include reports of harmful biologic effects of extremely low level ionizing radiation; the carcinogenic effects of electromagnetic fields associated with power lines; recent reports about cellular phones as a cause of brain tumors, etc. How long will it take for these examples of Pathologic Science to be dealt with appropriately?

In summary, concerns with ethics in research go beyond simply dealing with the elimination of outright fraud. The relevance of the other problems I described should be considered.

Steven Weinberg

I was moved by what Dr. Lee had to say about the standards of good behavior in scientific research. As it happens, I went about preparing my talk from the other direction, thinking about bad behavior in scientific research. For this purpose I naturally turned to an authoritative list of varieties of bad behavior, known as the Seven Deadly Sins. They are listed as: pride, covetousness, lust, anger, gluttony, envy, and sloth. I only have a short period for my talk so I will leave lust, anger, and gluttony to the afternoon panels.

Under the heading of covetousness, envy, and sloth, can I suppose be found the motives for the varieties of scientific misbehavior which are most often discussed in magazines and newspapers. For instance, the National Academy of Sciences has listed varieties of scientific misconduct the fabrication and falsification of data and plagiarism. I regard these as minor evils because, based on my own experience as a physicist, they play a minor role in the history of science. I am not aware of any serious deliberate case of falsification or fabrication of data in physics in this century. (Certainly some wishful thinking, but I believe sincere wishful thinking.) We read about cases of falsification and fabrication of data in other sciences, which might suggest either that physicists are morally superior to other scientists, or, on the other hand, that what we do is of so little importance, that there would be no motive for fabricating or falsifying data in physics. I am disinclined to accept either theory, so from my experience in physics, I am led to conclude the importance of falsification and fabrication of data in science is exaggerated, partly for political reasons and partly because of a pervasive hostility to science which is one of the problems of our society.

But there are varieties of misconduct that I think are worth considering, though they are rarely discussed in public and, in fact, I think not to be discussed elsewhere in this Forum. These evils come under the heading of the seventh sin, the sin of pride, which as you may recall is traditionally regarded as the most awful of the sins, and the one that created career problems for Lucifer.

Doing good work in science very naturally fills us with pride. It is not surprising, therefore, that some of the greatest scientists, as the years pass and as their honors and distinctions gather around them, begin to fall under the illusion that they can go on doing great science relying on their own unaided intellect, even though they lose sympathy with and get out of touch with the work of younger and less famous scientists. We can think of Einstein, whose achievements of course are incomparable, but who in the last twenty years of his life did nothing of great importance because, I believe, he was out of touch and out of sympathy with the work on quantum physics going on around him. Paul Dirac's hostility to modern quantum field theory provides another example, more familiar to physicists than to the public, and I'm sure there are many examples outside of physics. (Some living physicists provide similar examples, but in this talk I shall follow the rule of speaking ill only of the dead.) I hasten to say that I do not put the cases of Einstein and Dirac and some others in the category of scientific misconduct, because these physicists never tried to impose their views on the scientific community, never tried to arrange that only research in the directions that they were interested in would be pursued. Pride of this sort really begins to hurt progress of science in cases where the scientist becomes a mandarin, convinced that his work is the only sort that need be pursued and that his point of view is the only one that is legitimate, and then tries to impose this view on the scientific establishment.

Though it is based only on anecdotal evidence, I have a strong impression that this is the role that was played by Werner Heisenberg in the last twenty years of his life. In the late 1950s, he developed a nonlinear quantum field theory, that was supposed to describe all forces and particles. I heard Heisenberg describe this theory in a special seminar when I was a new instructor at Columbia. Even then, it seemed to me to be devoid of any rationale and also impossible to solve, and therefore not worth working on. Most other physicists felt the same way, and ignored Heisenberg's theory. It was a time of great excitement in particle physics, including the discovery of parity nonconservation, the two-component neutrino, the "V minus A" universal weak interaction, and so on, but Heisenberg came to the conclusion that the progress of physics depended on seeking solutions of his field equations, and he did much to impose his view on physics institutes in Germany. He almost succeeded in excluding mainstream quantum field theory and elementary particle physics from German universities and I believe that he did delay CERN, the great experimental laboratory in Geneva, for several years, because he felt it was no longer necessary to do experiments on elementary particles, but only to solve his equations.

There are other examples. I believe that DeBroglie and, earlier, Poincaré, played a similar role in France. Perhaps Sakata did at Nagoya. I can't speak about it authoritatively; it is a topic that needs further study by historians. I'm glad to see that a graduate student in the history of science at Harvard, Catherine Parson, is taking up the case of Heisenberg and trying to document the extent to which he impeded the progress of physics in Germany after 1950.

I think it is striking that the examples I've given here have all been outside the United States. I don't know of any American physicist that has succeeded in really distorting the direction of scientific research by imposing a narrow personal view that was out of touch with the main stream of research. I think this is to be attributed to the great disorganization and decentralization which is the glory of American academic and scientific life. We have many research universities. We do not have a clear ordering of the relative prestige of these universities. Some of our universities are public, some private. It would be difficult to imagine any one of them dominating the whole of American science. Even within a single physics department, it would be very difficult to see how anyone could dominate the research of all the professors, since our departments are run by chairmen who take the job generally as an unpleasant public duty rather than as a tribute to their eminence and prestige.

I hope that we will always hold to our apparently inefficient and disorganized American system of scientific research. We do not need congressional committees to help us to avoid obstacles to scientific progress caused by the sort of mandarinism that I have been discussing; our own peculiar American higgelty-piggelty system is doing very well. I would certainly urge our policy makers to avoid any rationalization of the system of scientific support. I think that it is quite wonderful that it is not possible to explain in rational terms which agency is supposed to support which branch of science, because this provides a tremendous protection against mandarinism.

I thought I would end here, but, in listening to Dr. Bugliarello's talk, I was reminded of another kind of pride. I thought it would be appropriate to mention it here because it presents a present danger to my own field. Dr. Bugliarello spoke of a final theory and the end of science. It is true that we elementary particle physicists are proud of the particular historical mission that we see as the motivation for our work. We have seen through the centuries that the increasing explanatory power of science allows more and more to be explained in terms of fewer and fewer fundamental principles. Suppose you ask any question about the world as we see it — for instance, why is the sky blue? We can explain this in terms of the scattering of light by dust particles. And then if you ask why light scattering works the way it does, we can explain that in terms of Maxwell's equations. And if you ask why these equations are true, you can get an answer in terms of the requirements of relativity and quantum mechanics for spinning massless particles. Wherever we are not blocked by the interposition of historical accidents, which are so important in biology and geology, wherever we can trace a chain of explanation back sufficiently far, we come to an answer in the theory of elementary particles. Some of us elementary particle physicists, perhaps most of us, have taken it as our aim to carry this further to a final theory that which will be at the end of all chains of explanation. I do believe in a final theory, and have devoted my own working life to pursuing it, but I don't believe in an end of science. I think that a final theory of the sort that I'm talking about will end the search for those explanatory principles that do not have deeper explanations, but it won't do anything to solve the other great outstanding problems of science, problems of thought, of how life started, of the evolution of the earth and galaxies. In this sense, there will be no end of science; it will go on forever.

Nevertheless, our talk in elementary particle physics of seeking a final theory somehow offends the pride of some of you in other branches of science who are pursuing these other grand questions. So I would like to appeal to my fellow scientists here in this room that, while you are pursuing your own branches of science, which will, as I think, go on forever and which will be endlessly fascinating, please don't turn your backs on the efforts of elementary particle physicists, or at what, after all, has all along been one of the nobler aims of science: to trace to their roots the explanations of why the world is the way it is.

Managing Competing Interests: Chastity vs. Promiscuity

Carl Djerassi

Why use such a sexually charged subtitle at a symposium dealing with ethics and values in science? Is it just to attract more readers or listeners? Might not "Perception is Reality" be just as suitable for a talk dealing with potential conflicts of interest? While I welcome any extra attention or attendance stimulated by an ambiguous title, for the purpose of this presentation, I equate chastity to operating within the confines of academe, whereas promiscuity encompasses the "chaste" academic's involvement with the tougher world of industry and business. In other words, I use the broader dictionary definitions of chastity (**purity in conduct and intention**) and of promiscuity (**mingling of persons or things**) to examine some of the conflicts of interest facing today's academic scientist.

In theory as well as in many religions, chastity and promiscuity are judged in absolute, moralistic terms. In reality, they are invariably examined subjectively, accompanied by a heavy dose of sanctimoniousness. Given the complexity of these issues, which can hardly be treated adequately within the time allotted to me, and given my current autobiographical frame of mind (having published no less than two autobiographies[1,2] within a couple of years), I will examine the topic almost entirely by way of personal experiences — largely promiscuous ones — covering a span of half a century as a research scientist, before trying to draw some generalizations. Right at the outset, I shall puncture any bubble of suspense by announcing my intention to demonstrate that professional and intellectual promiscuity — as defined by me — need not necessarily be unethical. In fact, it can contribute to the promise of science and hence of society.

The first seven years of my post-Ph.D. life, all spent in industry, were professionally chaste and monogamous. They were also extraordinarily exciting, considering that they included the first synthesis of a steroid oral contraceptive as well as the first successful synthesis of cortisone from a plant raw material — all accomplished in the scientific wilderness of Mexico.

The following five years as a chemistry professor at Wayne State University still fell more or less within that definition of chaste, professionally monogamous behavior since I followed standards of academic conduct totally acceptable in the 1950s: I was a full-time professor doing occasional consulting for industry (remunerated at then existing levels of piddling compensation) within the strict time confines imposed by university regulations and by a self-imposed standard of never canceling classes. But in 1957, twelve years after leaving graduate school, professional bigamy became my modus operandi.

In that year, my research group at Wayne State University, consisting of ten postdoctorate fellows and eight graduate students, pursued research in the fields of natural products chemistry and of optical rotatory dispersion — all supported financially by the National Institutes of Health, the Rockefeller Foundation, the American Heart Association, and the National Science Foundation. For medical as well as personal reasons, I took a two-year leave of absence from my professorship at Wayne to serve as vice president for research of Syntex in Mexico City — a corporation for which I had worked full-time from 1949-1952 and for whom I had consulted in the intervening years. During that Mexican interval, I maintained my academic research by frequent telephone contacts, extensive correspondence (in pre-FAX days), and bimonthly visits to Detroit. The Syntex research directed by me in Mexico City focused entirely on the development of new steroidal medicinal agents (corticosteroids, anabolics, oral progestins, etc.) which had nothing to do with my concurrent academic research projects in Detroit funded by governmental or philanthropic organizations. The bulk of that industrial research was published in the scientific literature, but it also resulted in the filing of numerous patent applications and the issuance of dozens of U.S. and foreign patents.

My two lives were almost separate, but not totally so. In my capacity as research vice president of Syntex, I persuaded the company to underwrite an ambitious, in-house postdoctorate fellowship program — probably the first to be organized by a pharmaceutical company. Of the first ten foreign incumbents to come to Mexico City, all but three were former graduate students or post-doctorate fellows from my Wayne research group. Five of them (two from the U.S., the rest from the U.K., Costa Rica, and Italy) eventually accepted full-time employment with Syntex and rose to the highest levels of Syntex management — one eventually becoming its chairman and CEO. I dwell on these details, because these five individuals would not have ended up at Syntex had it not been for my beginning bigamous professional life.

In 1959, when I was about to resume my academic position at Wayne State University, I was offered a professorship at Stanford University, which I accepted. Within a couple of years of my arrival at Stanford University, the first storm cloud about an apparent conflict of interest arose. The NIH had just introduced the requirement of annual invention disclosures: was any of the research undertaken under NIH auspices patentable? Just as in Detroit, the National Institutes of Health had continued to be the main source of my academic research funding in California (I do not recall ever having solicited industrial research support during my entire Stanford academic career encompassing 33 years) and in my report to the NIH I listed no patentable inventions emanating from the chemical research of my academic laboratory group then amounting to approximately 20 graduate students and post-doctorate fellows.

"Aha! Gotcha!" I visualized a gloating NIH auditor exclaiming when I received a missive asking how I could possibly be publishing so many scientific papers from Stanford without filing a single invention statement with the NIH, whereas at the same time inspection of *Chemical Abstracts* demonstrated that I was the inventor of numerous patents, all of them assigned to Syntex.

There were two answers I could have given, but didn't: First, as a matter of principle, I had decided not to file any patent applications on any of my academic research results. Prompt publication in the scientific literature would prevent anyone else from patenting such work, which would thus be available without restriction to the general public. Second, I did not believe that the type of academic research performed by my Stanford group on the structure elucidation of natural products, and on the development of chiroptical and mass spectrometric methods was truly patentable.

Rather, I responded by referring to the record: over the course of three years, I had been simultaneously employed by industry (Syntex) — a fact that I had disclosed openly to the NIH when I went on academic leave from Wayne State University in order to be certain that my academic research support could continue — and that the various issued patents were all based on applications filed during my Syntex employment, covering solely research conducted at Syntex.

My explanation was not accepted at face value. Rather, the NIH froze all of my grants until some independent confirmation of my explanation could be secured. A consultation with Stanford University's outside legal counsel provided a chilling estimate of several years for the resolution of this grant embargo — an estimate that seemed almost conservative in the light of the fact that by then I had published over 400 scientific papers that the NIH auditor considered relevant. "Somebody other than you," Stanford's counsel opined, "will have to go through every one of those papers and through your hundred odd patents to certify that none of the latter relate in any way to your NIH-supported research. And that 'somebody' better be a recognized expert in steroid chemistry before the NIH will be swayed to lift the embargo."

"You can't be serious," I wanted to reply, but one look at the lawyer — who, after all, was supposed to represent *my* interests — convinced me that I better swallow such commentary. Were it not for the extraordinary generosity of our department chairman, William S. Johnson, an internationally recognized authority in the steroid field, who offered to undertake the monumental and thankless task of confirming the veracity of my response by going through all of my published papers and patents, I might well have been faced by academic scientific bankruptcy. There is one lesson I should have learned from that experience but didn't: *publish less*! The lesson that did stick with me for the next three decades of my professional life was the conviction that my unswerving decision to keep my academic and industrial research programs separate was a wise decision.

I have never before disclosed in public this contretemps with the NIH in the 1960s. But it would be naive to assume that only an eager beaver NIH auditor would discount the self-imposed, severe ethical standards of my increasingly bigamous professional life. Before disclosing more of the latter, and deriving some lessons relevant to this Sigma Xi symposium, I shall describe another incidence illustrating the attitude of my academic peers by quoting from my recently published autobiography:[2]

> Three years prior to its demise in 1980, the *Berkeley Barb*, an acerbic muckraking tabloid, published a long article criticizing the financial gains that had accrued to various university professors as a result of their association with the many biotechnology firms that had started to flourish in the San Francisco Bay area and around Boston in the shadows of Harvard and MIT.... The reporter quoted an apparently uncontaminated Berkeley professor to the effect that my academic position "hadn't kept Stanford chemist Carl Djerassi from privately patenting birth control steroids he discovered under his own name for profit, even though he had discovered them while doing NIH-funded research. Perhaps significantly, Djerassi...used his own company to market such steroids."
>
> I was not a reader of the *Berkeley Barb*, but several copies of this particular issue promptly landed on my desk. Since their allegation, that I used government funds to feather my personal nest or that of my industrial employer, could and should have had a major impact on my academic career and on any further government funding of my academic research, I responded immediately. I pointed to the public record, showing that the patent application on the oral contraceptive was filed in November 1951, that the patent was assigned to my then-employer Syntex, that my Stanford University affiliation had started only in 1959, and that I had not filed a single patent application since that time. I also added that I had never received any royalties for my work on oral contraceptives or for any of the other one hundred-odd patents of which I was an inventor while employed full-time by industry. Though the *Berkeley Barb* was not known as a paper likely to print retractions, in this instance they published a full-page palinode.
>
> My reason for telling this story is that while the reporter printed an unequivocal *mea culpa* for not having checked the public record or interviewed me, he did insist that he had both quoted the Berkeley professor correctly and been given the impression "that Dr. Djerassi's alleged private patents on birth control drugs were

common knowledge in the scientific community." In that respect, I believe the reporter to have been dead right. There is little I can do about that perception, which is caused by a mixture of academic naivete and wishful thinking, often also tainted by professional jealousy.

When I resumed my full-time professorial career following the end of my leave of absence in Mexico City, I remained a member of Syntex's small board of directors — a responsibility I could easily fulfill within Stanford's limitations for outside activities of its academic staff. I would now like to illustrate how I was able to use the one-day-per-week allowed for "professional promiscuity" for a variety of entrepreneurial activities, which not only did no harm to my academic activities or graduate students, but actually produced major societal benefits: several novel useful products as well as a variety of new employment opportunities for technically trained personnel. With some condensation, I quote again from my autobiography:[2]

Shortly after my arrival in Palo Alto from Mexico City in the early autumn of 1960, I persuaded my fellow Syntex directors that the time was ripe for diversification beyond steroids. My first Stanford friend, Joshua Lederberg, had been awarded the Nobel Prize in Medicine in 1958 for the discovery of bacterial genetics; Arthur Kornberg, chairman of Stanford's new biochemistry department, had won it the following year for his enzymatic synthesis of DNA: almost overnight, Stanford had become a world center in the new field of molecular biology. Since no pharmaceutical company had as yet made a significant commitment to that area, I suggested that Syntex be among the first. Within a year, the Syntex Institute for Molecular Biology was established in a new one-story building on the Stanford Industrial Park — with me its head as a part-time Syntex vice president, and Joshua Lederberg as its advisory research director — both of us performing these duties within the one-day-per-week private time allowed full-time Stanford professors. Lederberg was chiefly responsible for setting our scientific priorities and hiring the research group leaders. (Fred Terman, who, as provost, had brought me to Stanford, proudly participated in the 1962 inauguration of our institute because he saw his dream of attracting biomedical industry to the Stanford Industrial Park realized in record time).

In early 1963, when Syntex felt ready to enter the American pharmaceutical market under its own name with some of the drugs we had invented in the late 1950s in Mexico, I urged that the company establish its U.S. headquarters next to Stanford — where we already had our Institute of Molecular Biology, and would be next door to a major medical school. The clincher was my argument that, since virtually all American pharmaceutical companies were east of the Mississippi, we would have no competition in attracting top scientists to the San Francisco Bay area.

Within a couple of years, the molecular biology group shifted to the new Syntex research complex, fortuitously vacating the one-story laboratory building that had been its first home. Timing could not have been better. William Little, a Stanford physics professor, who had just published a controversial theory of superconductivity, had approached me with a challenge. Until then, superconductivity had been displayed only by some carefully purified metals near absolute zero, which required the use of expensive liquid helium. Little proposed that certain hypothetical *organic* polymers should also be capable of superconductivity — and at near room temperatures. But to prove Little's hypothesis, one would have to synthesize a polymer that had never been seen before: a linear, conductive, polymeric backbone, with branching dye molecules bearing electric charges. Such work needed participation by industrial experts conversant with organic synthesis and with practical applications in electronics and possessing deep pockets. I recommended to the ever-adventurous Syntex board that we undertake the backing for Little's superconductivity gamble by forming a joint venture with Varian Associates (along with Hewlett-Packard, one of the first tenants of the Stanford Industrial Park).

In just a few months, Synvar Associates, as the new partnership was christened [subsequently abbreviated to Syva], was housed in the newly vacated Institute of Molecular Biology building. Since the building was ten minutes walking distance from the headquarters of the two corporate parents, Syntex and Varian, it was easy for the board of governors to meet almost daily at lunchtime. I served as chairman and undertook the responsibility for locating and hiring the key staff members, starting with Edwin Ullman, a Harvard Ph.D., as scientific director.

Though skeptical about the possibility of actually synthesizing Little's designer molecule for organic superconductivity, Ullman was willing to give it a try. He would, after all, have as a consultant my Stanford colleague Harden McConnell, a chemical physicist, who himself had a theory of superconductivity (based on certain types of metal sandwiches). In addition, McConnell was also studying the biophysical properties of stable free radicals whose electron deficiency made it possible to detect them in complex mixtures by electron-spin resonance (ESR), for which Varian was the principal supplier of instruments. We offered Ullman the opportunity of hedging our research bet by pursuing McConnell's idea in parallel with Little's project, and to see whether such stable free radicals might actually have practical utility.

Consciously or not, any enterprise subsidizing a group of bright and adventurous researchers becomes a proponent of serendipity, because such a venture is likely to pay off sooner or later, even if not in the area originally envisioned. In 1971, Syva abandoned the attempt to synthesize Little's dream polymer, because we had begun, in the early 1970s, to strike gold in our search for practical applications of organic free radicals.

Two decades later, the company's annual revenue from diagnostic products arising from the original stable free-radical work had passed the $200-million mark. The seminal idea had been provided by Avram Goldstein, then head of the department of pharmacology at the Stanford Medical School and a consultant to Syva on possible biological applications of the unique ESR properties of stable free radicals. A neuropharmacologist of international repute, Goldstein was particularly interested in opiate addiction and called to our attention the need, in methadone treatment centers, for a fast and sensitive method to screen patients who might be taking heroin surreptitiously. He also contributed to the Syva team's invention in 1970 of the FRAT (Free Radical Assay Technique) approach to the detection in urine of traces of morphine, a metabolite of heroin. Varian made a couple of prototype modifications of their research ESR machine to provide clinical laboratories with a simple "black box" which would convert the free radical's ESR signals into a graphic output indicating the presence and amount of morphine in urine. From this point on, developments proceeded at a stunning rate.

Within a year, President Nixon announced the initiation of a compulsory urine analysis program (based on Syva's FRAT technique) designed to detect opiate drug abuse among servicemen in Vietnam. The true extent of such abuse was unknown, but there was great worry that we would be spreading the epidemic at home if we returned soldiers who were actively addicted and physically dependent on opiates. Our first reagent order from the army amounted to nearly $2 million, which transformed us overnight from a research venture into a business. Syva became a household word in clinical laboratories when Ullman's team developed a second method, termed EMIT for Enzyme Multiplied Immunoassay Technique, for detecting drugs of abuse, as well as many other therapeutically significant medicines. The rationale for this second application is easily appreciated: since in the treatment of many chronic diseases — for instance epilepsy, asthma, heart problems — it is important to tailor the dosage to the individual patient, rapid and simple assays are required for these particular medicines. EMIT (*Time* spelled backward) proved to be ideal for such purposes.

Syva's scientific and commercial success is a first-class example of the synergy generated when the interplay between academia and industry is allowed to proceed in an enlightened environment — especially when the academics serve not just as consultants, but also as initiators of research projects. (Generally, this works best in small entrepreneurial settings and not in large establishments, where Parkinson's Law is inexorably operating on a grand scale.) Little, McConnell, and Goldstein divided their one-day-per-week "free" day into four or five lunch periods at Syva, which offered them almost daily opportunities with Syva scientists. These highly focused contacts, undiluted by administrative trivia or telephone interruptions, produced intellectual sparks that benefited both constituencies. But there is no question in my mind that the professorial participation was the indispensable component to Syva's success: Little's superconductivity theory had been the raison d'être for the enterprise; McConnell's free radicals had provided the diversification; and Goldstein had pointed the way to the first practical, biomedical applications. [For the purposes of my argument, it is important to note that these three academics had not only provided the initial, intellectual research lead, but — contrary to the usual role of academics — they had also participated intimately in the work leading to practical applications. They may have lost their academic chastity, but in the process they had become more realistic and wiser academics.]

My own corporate activities really exploded in 1968. Alejandro Zaffaroni, until then the President of Syntex Research and of the company's U.S. commercial branch, had decided to leave Syntex to form (on the Stanford Industrial Park) his own company, ALZA, dedicated to developing not more potent or specific drugs but methods of delivery. I assumed the post of president of Syntex Research (while retaining the chairmanship of Syva's board of governors) and formally went on a half-time academic schedule (with the approval — indeed, encouragement — of Stanford's provost, Frederick Terman). The operative term is *formally*, because in actual fact I reduced neither my academic research load nor the size of my research group. Even my teaching schedule at that time was not much lower than that of my full-time colleagues. I was able to accomplish that by tight scheduling and because my Syntex office was only a 10-minute drive from my Stanford office. In general, I wore my professorial hat from 8:00 to 11:00 AM and from 3:00 PM onward, whereas the middle four hours of each day were spent running the Syntex research enterprise.

One of the first steps I took as president was to escalate Syntex's efforts in the insect field. Syntex's interest in insects dated from the mid-1960s, when the insect molting hormone ecdysone was shown to be a steroid. For its mode of action to be studied in any detail, larger amounts would have to become available, and a Syntex group headed by John Siddall, a British postdoctoral fellow, was among the first to publish a successful ecdysone synthesis.

Just as we started to gain some knowledge of insect physiology by establishing a consulting relationship with the Harvard insect biologist Carroll M. Wilson, a second endocrinological bombshell exploded. Herbert Röller, then at the University of Wisconsin, announced the successful isolation, structure elucidation, and synthesis of an insect hormone that governs certain processes peculiar to the early developmental stages of insects, notably, the larval phase. Only when production of this hormone stops, can the insect mature and reproduce. Carroll Williams, himself active in this field, had called attention to the potential of this hormone as a new method of insect birth control. What was so exciting about Röller's announcement was that the structure of his "juvenile hormone" is so much simpler than that of ecdysone that one could conceive of devising a sufficiently economic synthesis of it for use in practical insect control. As far as we at Syntex were concerned, the new science of insect hormones seemed just the key to an environmentally more benign alternative to conventional insecticides, and we decided to find it with the help of two academics — Carroll Williams and especially Herbert Röller (who soon became infected by the same professional promiscuity virus to which I had succumbed earlier).

Even if the science went right, we estimated it would take at least $10 million and five years for such a research effort. In 1968, Syntex was still too small for such a gamble: the company's entire annual research and development budget was on the order of $10 million. We decided to put all of our patents, know-how, and key research personnel from our insect research into a separate company, named Zoecon Corporation, in which Syntex would retain 49-percent ownership. The remaining 51 percent would be spun off to Syntex stockholders as a stock-rights offering for $10 million.

As if I didn't have enough to do at Syntex, Syva, and Stanford, I now became also president and chairman of the board of Zoecon, which was housed in the Stanford Industrial Park. Herbert Röller, who had just moved from the University of Wisconsin to a professorship at Texas A & M, served as Zoecon's part-time vice president in charge of research — an example of academic promiscuity conducted over a much greater distance than my own. His principal functions were attracting the key scientists for our biological laboratories, establishing a top-notch insectary, and supervising by long distance and periodic short visits the progress of the biological program. John Siddall, who had been the key Syntex chemist on the ecdysone and juvenile hormone projects, together with a former Syntex postdoctorate fellow from Australia, Clive Henrick, headed the chemical program. Two other young scientists who eventually filled key functions at Zoecon were my former Stanford Ph.D. students, John Diekman (who joined Zoecon's budding development and registration department and eventually ended up as president of the company) and David Schooley (who became the head of the insect biochemistry division and is now a professor of biochemistry at the University of Nevada).

We set out to revolutionize the insect control field by attempting to base our approach on Röller's insect juvenile hormone. All one has to do to control a particular insect population, we reasoned, is expose the insect continuously to its own juvenile hormone, so that it can never mature and replicate. (The human counterpart to such a form of birth control would be the administration to infants of a chemical preventing the onset of puberty: although the children would never become parents, they might well survive beyond the usual time of puberty and thus certainly be costly in economic and social terms). Even this scenario was hypothetical, because the natural juvenile hormone, when administered by conventional methods such as spraying, was quite unstable. Both bacterial enzymes and sunlight decompose the natural hormone so quickly that its half-life under field conditions probably does not exceed a day or two. Administration would have to be repeated so frequently as to be prohibitively costly. We had to hope that chemical alteration of the insect juvenile hormone might create a more active variant that might be more stable under field conditions — not unlike what we had accomplished nearly twenty years ago at Syntex in Mexico City by synthesizing a more effective, orally active, congener of the natural female hormone progesterone, thus opening up the field of practical oral contraception in humans.

We decided to focus on public health pests — mosquitoes, fleas, flies, fire ants — that are harmless as juveniles and become dangerous or annoying as biting, stinging, or blood-sucking adults. In the 1960s and early 1970s, malaria was still the biggest killer worldwide; and the responsible vectors—the various mosquito species — were then primarily controlled by DDT. Now that DDT was being banned in many parts of the world, and some mosquito species had started to develop resistance to it and other conventional insecticides, it seemed to us that the world was ready for a new approach to mosquito control. Various mosquito *Aedes* and *Anopheles* species headed our list of target insects, followed by flies and certain beetles.

In less than eighteen months, Siddall's chemical group synthesized several hundred variants of the natural juvenile hormone and finally arrived at a structural analog 2,430 times more active than the natural hormone in a mosquito assay and less prone to bacterial or photochemical decomposition in outdoor, aquatic environments. We christened the new compound ALTOSID (Palo Alto + John Siddall).

It took us some time to discover how little people are willing to pay for the wholesale control of public health pests at the breeding places; how little premium is placed upon preventive rather than acute control. Most people are willing to do something about mosquitoes that are biting them; fewer are willing to take seriously future generations of mosquitoes that may never appear. Our highly biodegradable product has to be administered at the correct time, as a persistent pesticide, like DDT, does not. Furthermore, for mosquito control by means of ALTOSID to be effective, it must be carried out on a large scale, typically by mosquito-abatement districts or other public agencies. There would be little purpose in applying the material to a pond in one's backyard if the neighboring swamp is not treated.

To gain public attention, we felt that we had to demonstrate the practicality of hormone-promoted insect control with some concrete example, and that if we could accomplish this in an aquatic environment, all future applications would have an easier passage through the bureaucratic maze of the EPA.

In fact, one of the most persuasive indications of ALTOSID's safety margin was the World Health Organization's eventual recommendation that our product could be added to human drinking water — a feature that proved useful for malaria control in countries like Thailand, where drinking water is frequently stored in open vessels that are potential mosquito breeding grounds.

Concurrently with our insect hormone research, we undertook a second R&D program in insect pheromones, and especially sex attractants, many of which had only recently been isolated and synthesized. Indeed, by the early 1970s, we had become the largest and most diverse supplier of pheromones in the world. Still, Zoecon would never have emerged from its pupal shell if the company had depended solely on marketing ALTOSID for mosquito control and pheromones for insect monitoring. Our early field trials, however, had taught us a lot, and I became convinced that Zoecon would become a viable adult company that would pioneer fundamentally new approaches to insect control. Around that time, coincidental with my increasing concern about the societal ramifications of my industrial life, I had also come to realize the depth of my belief that small is more beautiful than big. My metamorphosis into super administrator seemed inevitable, given the track my life was then on. It had become obvious to me that soon there would not be enough hours in a day to continue, in addition to my Stanford professorship, as president of Syntex Research as well as the chairman and CEO, respectively, of two growing industrial ventures — Syva and Zoecon. Syntex itself was then expanding so rapidly that everyone expected me to relinquish my other corporate interests and to concentrate on it. But I chose the youngest company — Zoecon — for my nonacademic hours. I fantasized about duplicating the Syntex experiment once more — converting a small, innovative research enterprise into an integrated operation involving research and development, manufacture and sales — and to do this in a field of high societal benefit. I decided to stake my industrial career on this conviction and on 31 December 1972, I severed all of my connections with Syntex.

By 1976, Zoecon's position in the insect field mirrored that of Syntex twenty years earlier in pharmaceuticals: we were hardly known by farmers or the public; but in scientific and industrial circles, we had become known internationally for the quality and quantity of our publications on insect hormones and pheromones. As a form of indirect compliment to our increasing reputation, two years later, the giant Occidental Petroleum Corporation acquired Zoecon by making an extremely attractive offer to our stockholders. For five years, we operated with remarkable independence within the Occidental Petroleum framework, with me continuing in my usual part-time mode as CEO of the Zoecon subsidiary. During that period, our new corporate parent increased our research budget three-fold, thus permitting us to become one of the first agrichemical organizations to establish a molecular biology unit. In 1983, by which time Zoecon's annual sales had crossed the $100 million benchmark, Occidental Petroleum sold its Zoecon subsidiary to the Swiss pharmaceutical giant, Sandoz, Ltd. The subsequent history of Zoecon is not relevant to the subject of this article and, in any event, has already been told in detail in my autobiography.[2]

What conclusions can be drawn from this personal account of a scientific career flitting back and forth daily between the academic and industrial worlds, a seeming mine field of potential conflicts? And what are the perceived sources of those conflicts, remembering that perception is frequently synonymous with reality? Is it the corrupting influence of money? There is no question that money — in the form of outside income, whatever its source, beyond one's academic salary — plays a role, but how should it be judged? The professor earning a couple of thousand dollars of royalties over the course of several years from the publication of a monograph that took years to write would clearly be exempt from such criticism. What about the author of a textbook that eventually earns the professor royalties in the six or seven figures? Such mega-royalties are not unheard of in popular texts, yet writing such texts in one's academic specialty is rarely considered a conflict.

Few hackles are raised when an academic consults for industry within the generally accepted one-day-per-week personal time, provided the retainer is modest. But what if five-digit sums or stock options are received? Clearly, red lights go on immediately, because of suspicion — partially fueled by professional jealousy — that such levels of compensation are bound to contaminate the academic purity of the recipient, especially so when the research ideas or results of the academic scientist are exploited in entrepreneurial settings. In one way or another, most academic research can be traced back to a stage where the public coffers (through the NIH, NSF, or other agency) supported some of the work. Substantial financial rewards, especially in the form of equity, are then almost always equated with private enrichment from the taxpayer's pocket without ever considering the huge indirect benefits frequently accruing to society.

But is money the only source of such conflicts of interest? Chauncey Starr[3] succinctly covers other dangers only too pervasive in academe:

> "The academic researcher seeks the accolades of his professional colleagues, and depends on such recognition for promotion, tenure, professional awards, and grants from government agencies and foundations. The politics of science reveals the extent of the manipulations by individual scientists to achieve these rewards. For most academics, these are more persuasive inducements than the money flow from business affiliations. The recent flurry of investigations into the ethics of scientists discloses the greater value placed on these non-monetary goals."

And what about the troubling question of the perceived exploitation of students and co-workers under such circumstances? We all know of the extent to which graduate students and postdoctoral fellows are pressured into accepting 60- to 80-hour macho work weeks for the furtherance of the professor's reputation. We tolerate such visible, daily exploitation, but when that professor also grazes in an industrial pasture, questions of impropriety are invoked, which are as applicable to purebreds in the academic corral. To paraphrase one of Koshland's recent *Science* editorials,[4] perception of potential conflicts of interest is directly related to the degree of sanctimoniousness displayed by the critic.

My highly abbreviated list of actual or perceived conflicts demonstrates how gray these black and white problems really are. As illustrated by my auto-biographical record, I have experienced virtually all of them. In the 1960s and even early 1970s, entrepreneurial excursions of academic scientists into the industrial realm were relatively rare, and mine were more an exception than the rule. But nowadays, it is almost impossible to find a leading scientist in most cutting-edge scientific disciplines — especially in biomedical or material science areas — whose involvement with industrial enterprises does not carry monetary rewards (especially in terms of stock options or stock ownership) of a magnitude that invites instant suspicion and criticism.[5,6,7]

In my own instance, I have taken some safeguards to reduce the level of justified concern: I kept the subject matter of my academic research completely separate from my industrial research involvements and did not patent any academic research; my formal academic time (and salary) was reduced by half so as to raise no question about my having used "tax payer's time" for the benefit of the industrial enterprise of which I served concurrently as a part-time executive or member of the board of directors. But it is quite unrealistic to propose that most academic scientists desiring some significant relationship with an industrial entity restrict themselves in such manner. First, there is nothing illegal about patenting academic research. Indeed, from the perspective of societal benefit, such activity frequently expedites practical uses or even spawns them. Second, most scientists are not in a position, for practical or personal reasons, to separate their academic research from what they could offer concurrently to industry.

It is not possible to turn a clock backward. I have no way of predicting what else I would have accomplished in science, had I been professionally chaste. In theory, I would have had more time at my disposal for academic research, but would I have worked with the same intensity and efficiency if the pressure of time had not hung over me as a result of my non-academic commitments? I can be much more certain what would not have happened if I had remained professionally chaste in my academic ivory tower.

Syntex would almost certainly not have moved to California, but rather to New Jersey, when the corporate decision was made to establish an American home. While Syntex might also have flourished outside the exciting entrepreneurial culture of Silicon Valley, the Stanford Industrial Park would have lost one of its two largest employers. My entrepreneurial participation in the genesis of companies such as Syva and Zoecon may not have been essential, but it was crucial. Several thousand employees and a couple of hundred industrial postdoctoral fellows would have held these positions elsewhere or perhaps not at all. A fair number of senior colleagues, both at Stanford and elsewhere, were enticed by me to indulge in similar promiscuous behavior, and while none of them did so to the extent displayed by me, it did have an effect on their professional lives and that of several companies, in general to their mutual benefit. And from a wider societal perspective, it is impossible to gauge when, where, or even whether some of the important practical advances made available to the public by these companies would have occurred without such intimate participation by academics of the Goldstein, McConnell, and Röller class.

Stanford University has been a direct beneficiary — and not only by acquiring some important industrial lessees for its land — through major gifts by some of these corporations. My professional promiscuity made me a much more interesting and diverse teacher; my course syllabi included topics of social policy and practical relevance I would not have even thought of, had I not experienced them in that outside world.

I list these societal pluses while admitting that for my industrial and entrepreneurial activities, I was remunerated handsomely in terms of money and/or stock equity. As alluded to elsewhere,[2] a significant portion of those assets ended up in philanthropic ventures. But that does not change the perception, or even the fact, that this was remuneration for services rendered — even if much of that service was intellectual — and that it occurred on non-academic turf.

I believe that the great majority of academics with industrial connections are scrupulously honest, and that such associations benefit the academic and industrial constituencies. Nevertheless, it is indispensable that the nature of these extra-academic involvements and the resulting remuneration — notably in terms of stock options or stock ownership — be openly documented. In addition, some general rules of accepted behavior should be promulgated: the hours/week a full-time professor is permitted to dedicate to outside activities; what outside positions a professor may hold (e.g. director, officer, advisor); whether part-time professorships are permitted; what length of unpaid academic leave of absence is tolerated; how a university's patent policy is implemented with specific reference to royalties and to the participation of graduate students and other research personnel. These are only some of the major topics that may have to be addressed and which will differ in content, depending on the nature (private vs. public) and other special criteria of the academic or grant-giving institution.[7]

Disclosure and clear-cut, unambiguous guidelines are probably the best safeguards to prevent avoidable conflicts. What is undesirable is to micromanage the issue by trying to anticipate every possible source of conflict, thereby creating a thicket of absurdly bureaucratic rules. A typical example are the recently promulgated rules governing codes of conduct by NIH scientists with industry — an interaction the NIH is now trying to stimulate — ranging all the way to the implication that a tuna-fish sandwich offered by an industrialist compromises an NIH researcher and hence should be turned down.[8]

I end by claiming that while the academic water does indeed get muddled as the incidence and intensity of academic-industrial interactions increases, so does its nourishing quality for society. But care needs to be exercised in keeping out the dirt.

References

1. C. Djerassi, *Steroids Made It Possible*, American Chemical Society, Washington, D.C., 1990.

2. C. Djerassi, *The Pill, Pygmy Chimps, and Degas' Horse*, Basic Books, New York, 1992. (Paperback 1993).

3. C. Starr, "Is peer review unbiased?" *Nature*, **357**, 354 (1992).

4. D. E. Koshland, "Conflict of Interest Policy." *Science*, **257**, 595 (1992).

5. M. Barinaga, "Confusion on the Cutting Edge." *Science*, **257**, 616-619 (1992).

6. C. Anderson, "Genome Project Goes Commercial." *Science*, **259**, 300-302 (1993).

7. C. Anderson, "Hughes' Tough Stand on Industry Ties." *Science*, **259**, 884-886 (1993).

8. C. Anderson, "NIH Scientists chafe under ethics rules on industry ties." *Nature*, **357**, 180 (1992).

*The following five presentations are from a panel on the ethical issues seen
from the perspective of postdoctoral researchers. The panel was chaired
by Arthur L. Singer, Jr.*

Postdoctoral
Researchers:
A Panel

Arthur L. Singer, Jr.

I'm sure that one of the reasons why the steering committee meeting at
Wingspread decided a panel of postdoctoral fellows would be a good idea for
this forum was that postdoctoral fellowship activity is such a big deal in the
science establishment. For example, in 1991, there were nearly 25,000 Ph.D.
degrees awarded in science and engineering, including the social sciences.
About 70 percent of those Ph.D.s were awarded to males and about 30 percent
to females. It is interesting to note that the men and women with new Ph.D.s in
science in that year, and I think this holds true for the years surrounding 1991,
have gone on to postdoctoral study in approximately equal proportion. Nearly
40 percent of all new Ph.D.s, both men and women, go on to postdoctoral studies. Interestingly enough, breaking that down by fields, although men received
81 percent of those degrees in 1991 in physical sciences and 91 percent in
engineering — well above the 70 percent average — men and women went on
equally to postdoctoral study, or approximately so: 45 percent of men, 41 percent of women in physical sciences; in the life sciences, 57 percent of men and
54 percent of women. So postdoctoral study for newly minted Ph.D.s in our
science education system in the United States today is a very big deal. Thus the
planners thought that this piece of science education should be considered at
this forum.

I only have one other introductory comment before I introduce the four
postdoctoral research panelists. There was a survey done, admittedly very
small and consequently its findings can't be given great weight, at Bryn Mawr
College, reported at a subcommittee hearing in the House of Representatives in
1988. The survey covered about 245 scientific researchers. For postdoctoral

fellows it has some good news and some bad news. The good news is that 32 percent of those scientific researchers in the survey reported that they had suspected a colleague, a peer, of falsifying data or plagiarism. Only 14 percent, less than half the number that had suspected a colleague, had suspected an assistant or a student, a postdoc. That's the good news. The bad news is that of those who had suspicion of a colleague falsifying data or plagiarism, 50 percent — only half — took action to verify that suspicion. The other 50 percent did nothing in the case of a colleague. In cases of suspicion of an assistant or a postdoctoral student, 74 percent took action to verify the suspicion. So one lesson from that survey is that postdoc students are clearly less suspected than colleagues, but once suspected, postdocs, look out. Further inquiry into the suspicion is much more likely.

Garth Jones

Thank you for considering the viewpoints of postdocs and not just the people who hired them. There are an awful lot of postdocs out there in the world today. I would like to talk about the process we go through to get a postdoc, to find out what positions are available, who has the money, who is hiring, and then to ask the very, I think, serious question, is there any fairness in whatever system may exist and is there anything we can do that might improve the lot of people like me who just got a postdoc.

As time has gone on, the postdoc as an institution has become much more important. It seems that both academic institutions and industry that are hiring people want their new employees to have the breadth of experience that a postdoc provides. As the postdoc system has grown, you might think that there would be some way that people could go about systematically finding out how to go about getting these jobs. That's just not so. There are a number of ways you can identify who has a postdoc position available and if they might be interested in hiring you, but those are fairly much by word of mouth. Professor Djerassi, in a question earlier, referred to the chemistry community as a tribal culture. As one who is young in the tribe, it becomes difficult to break into it. If your major professor knows someone, their old major professor perhaps, who has positions open or is a particular friend, that can be very helpful. Some research groups end up having a tradition after awhile where a student will get their Ph.D. in one group and a significant proportion of those will move to the other group to do their postdocs, and vice versa. That's all well and good if you happen to be in one of those groups. If you are trying to break into one of those groups and you are from the outside that can be a little more difficult. So what a postdoc applicant ends up doing is writing unsolicited letters to a great number of chemists, as in my case, hoping for some kind of favorable response. You might think that there is, of course the trade journal, *C & E News*, that people would rather logically advertise for their positions. In my experience, there are really only three cases in which people advertise. If someone is a young, relatively unknown professor who has just received a major grant and needs a postdoc and no one knows to write them letters, then the professor can place an ad and perhaps will get some applicants. Some people advertise if they have funds that need to be spent right then, they don't have the time to wait for letters to come in. Third, and this is perhaps the most reasonable reason to advertise, if you're a chemist who wants to branch out into a new area and you want to find a new Ph.D. with a particular expertise that doesn't normally write you letters, you need to advertise. So as a result, most chemists at most universities, again, in my experience, receive dozens, up to several hundred, of letters a year, both from American chemists and from overseas, asking (in some cases begging) for any kind of favorable response. I think that generates a real problem, not only for the students who are simply picking names they are interested in and writing them letters, but also for these people who are deluged with letters, sometimes which they don't have the capacity to respond to.

One scientist told me a story over lunch once. He receives two or three hundred postdoctoral applications a year. He runs a large research group, but he isn't able to accommodate even a small fraction of these people who write him. One day he happened to have just received word that he got a major grant funded. He had a postdoctoral position open. So he simply took the post-doctoral letters he had received that day and looked through them. There were some very good applicants; there were some not so good applicants. He called up the best applicant and hired that person. He didn't go back through his files; he didn't have time to go back through his files. If he kept them all, he would have hundreds and hundreds of letters, probably a huge pile in his office. So I think that generates a certain unfairness in hiring. If you happen to be lucky enough to send the right letter to the right person at the right time and they decide they are interested in you, then maybe you can get a postdoctoral position. If you are either coming from a university or a group that is highly involved in this tribal club, it is much easier. I don't want to complain overmuch about status and universities and so on. I think of myself as one of the lucky ones. I came from a university with an admitted middling reputation and found a very satisfactory position at Stanford, but certainly I think that there are a lot of good chemists out there in the world who are being simply overlooked because they are a little on the outside of this culture. In fact, my position is with Professor Johnson at Stanford. He is an emeritus professor and at the time I was writing application letters, I was writing to people who interested me and while I was interested in his research, I had no way of knowing whether he was still research active and whether he was in a position to hire people. At the urging of my Ph.D. advisor, I decided that 29 cents for an additional stamp was a worthwhile investment, and here I am.

I want to bring up one additional aspect of this issue. One of the other panelists is my wife. She is also a chemist. We have the additional problem of having to locate postdoctoral positions hopefully in such a way where we could live together for the next year or two and that has been a real problem. We did not receive very much sympathy or understanding except in a couple of very dramatic cases where people went very, very far out of their way to help us, not to any avail, because the money wasn't there. We were told many, many times (and I think if I hear this phrase again I may scream): well, you may have to live apart for awhile. Well, I understand that and it certainly can be a reasonable thing for awhile (we have done it already) and are not willing to do it again except under the most extraordinary circumstances. So again, I can't complain too much for my own case because I'm here and my wife is here and it all worked out very well for us. But it seems that there are just endless possibilities for unfairness in hiring. We have no way of knowing how people look at applications. We have no way of knowing who needs people, and perhaps I don't want to propose a specific solution now. I think that is something that we all need to talk about. It certainly seems that there ought to be some way that is a little more organized for potential employers to find potential employees and match them up in some sort of systematic way.

Jan Gurley

My job in the next ten minutes is to try to give you a feel for what it is like to be a postdoc in a less than optimal situation. I have to say that I feel like the first speaker at an Alcoholics Anonymous meeting. "Hi, I'm Jan Gurley, I've been the victim of an authorship debate," or "Hi, I'm Jan Gurley, I'm from a dysfunctional lab." What I would like to do, besides introducing myself, is bring up a few very specific examples, what I call *Dragnet* cases. The stories you are about to hear are true, the names have been changed to protect the innocent, or not so innocent, whichever you may be. I have already noticed this is a very intelligent crowd. Several people have already said, "you're a physician, you're not a postdoc are you?" As though they're saying, "you know, you're an apple, you're not an orange, are you?" There is a definite, immutable, uncrossable line between these two groups. Luckily I have been a lab rat long enough to know that physicians probably aren't really scientists. But, luckily, this is also a very tactful audience and no one has pointed that out so far. One of the things that is really important is the reasons why someone who is a physician is talking about this. Part of the reason is that we couldn't find anybody else. That will be one of the issues also. It is pretty much thought to be professional suicide to stand up and talk about a bad experience that you've had. I'm going to speak about two different specific cases, bring out some of the points that these illustrate, and then conclude with some of the things that I think we should try and think about in all of this.

I spent two years doing basic science research in a laboratory in Boston. I have quite a few friends who have been living through various stages of their postdoctoral fellowships. One of the examples is of a person who was doing very well in a very productive lab. He had published a number of different papers, was working on a last project, and had applied for a number of jobs. Luckily he got one and was pretty much encouraged by his supervisor, his boss (which is how we refer to the immediate supervisor in these circumstances) to apply, was given great letters of recommendation, had a wonderful resume. But, when he accepted a job at a different institution, he felt a significant degree of backlash from his laboratory, who had assumed, because things had been going so well, he was just going to stay there. They had offered him not quite as good a position, to say the least. During the last five or six months that he was in this lab, he had one project that really needed to be finished that was a companion piece to a previous publication that was very well-received in a major journal. As time got shorter and shorter, the lab decided to invest more and more money in new techniques that would be quite good at speeding up the process of getting the results he wanted. As he decided to leave for his new job, he was told there wasn't enough time to get both the last few studies done or to write the paper. Which would he rather do. He said, well you know it would be impossible for me to retrain somebody in the time that's left to do just these few assays, why don't I do them and maybe the immediate boss could write up the final draft and they could publish it. About three months into his new job, without any of the information from his previous lab, he was called and told that he needed to write this paper. He explained to them that he no longer had the time, the information nor the resources to be able to do that in the position that he was in now. He was told that if he did not write the paper, he would not be an author. At this point he flew back to the institution where he used to work, tried on several occasions to meet with his immediate supervisor, who repeatedly told him he was too busy, took (because no one was really worried that he would do this) two of his lab books, initially under the theory that he would try to get the paper written at his new job and left, having never spoken to his supervisor. He then became quite angry about the situation and decided not to write the paper. The lab found out that he had the data and there was a stalemate that has never been resolved. Approximately a hundred thousand dollars of federally funded research was never published because the postdoc has the data and the lab wants him to write the paper and no one was willing to cross a line that was drawn.

The second example is of someone who was working in a laboratory and starting to look for a young, start-up kind of lab. He wanted a job where he thought he could make some significant headway on a problem. He specifically didn't want one of these huge, sort of "mill labs" (nothing personal, Dr. Djerassi) where there are 30 postdocs and you have your little piece of the pie and you are working on it. So he joined a lab that had just published a really important seminal article, and he was given this entire new (it's hard to say what I should call this to keep it totally anonymous) cell culture, cell line. This paper was published in the biggest journal it could possibly be in. In fact, their entire new allotment of lab space and grant was based on this one publication. It was given to him as his project to try and develop this, characterize it, and find applications for it. Over the next year and a half, working in incredible frustration with this cell line, he discovered it was not, in fact, what it was published to be. During the course of that year and a half, while he was doing this work, the lab had moved on to other things and had, in fact, made several other important discoveries and published several other publications that were regarded as equally important. The lab did not have its funding threatened from finding out that this cell culture was not what it was originally advertised to be. However, the fall-out from this person going to the lab and saying I think you were wrong, was so severe that he was ostracized from the rest of the lab. He was given no new projects to work on. He was not fired, but he was not given any work. He lingered in this limbo, becoming clinically depressed, for about three months, and being a fairly innovative person, decided to take his reagent to a different group and ask them what they thought it was, since it clearly wasn't what it was supposed to be. In the process of doing that, he found out it was an equally important discovery in a different field and something that this new lab had been looking for, for about 20 years. There was then a very difficult, ethical issue of trying to use this reagent, publish it, and say where it came from, when the first lab refused to print a retraction. It ended up where there is now in existence, in a very prestigious journal, a very awkwardly worded sentence about this substance, saying exactly where it came from and, unless you were aware of the situation, you would probably say this sounds really strange, but not really think about the implications. Luckily, this story has a happy ending. This person went on to do a lot of productive work, has his own grant, has a lot people working with him, and the thing that he discovered that wasn't what it was supposed to be, actually turned out to be even more important in a different field.

So what do these things tell us. These are just two examples. They point up several important issues that I think are common to all postdocs. The first is there are very few options available to postdocs when there is a true conflict. It seems like the only two that are really feasible are suicide or luck. That's pretty much it. You can have either social suicide by completely rebelling against your lab or you can have professional suicide. It's not always the postdoc who is right by any stretch of the imagination, but it seems like the suicide always works the same way. It's very rare that a lab is destroyed by a postdoc, where there are many postdocs who are destroyed by labs. What unfortunately is rare for all of us, is that there are very few circumstances, as in the second anecdote, where something that has become an irreconcilable situation has a fortuitous ending. Another point that this brings up is that the future is pretty much dependent upon your boss when you are a postdoc. The postdoctoral fellowship can be described in a spectrum of good to bad as either an apprenticeship, indentured servitude, or slavery, depending upon where you are. Hopefully it has the atmosphere of the trainee who is learning and producing in a good atmosphere.

The third important point is that you were right in the beginning — physicians really aren't the same, and not just because of research or new research. A physician, no matter what happens, is thought to be playing at what they are doing, because they always have a job and they always have an income, no matter how well things go. That is definitely not true for postdoctoral fellows. The issue here is power and dependency. Even though I started this as an Alcoholics Anonymous symposium, I'll try to avoid the dependency jargon. The fact is that it is not an equal relationship and when conflicts arise, the resolution can't usually be equally bartered.

Another important point is what I call the Calhoun, Georgia syndrome. Labs are small places. I'm from Calhoun, Georgia. It's so small that when a friend of mine lost my address, they just pulled off the interstate and said my name enough times and someone showed them where I lived. When you live in a small place, and believe me twenty people is a big lab, the types of repercussions for what you do are very similar to what happens in a small town. The worst thing that can happen to you is for you to be ostracized. You become a nonentity in a productive group. That is a type of intermediate area between suicide, firing, and success that is very common, unfortunately. It often times results in the syndrome that I'm sure all of us know of, people who are doing their third or fourth postdoctoral fellowship. Where they have gotten a job, it doesn't pan out, they don't get along, they have been labeled as, just like in a small town, the "loose" person, the "not so smart" person. Once that reputation has started, it's hard to change, and they end up trying to go somewhere else. The other issue that relates to the small town syndrome is that in small towns, people live there a long time. When you're a postdoc, you live there a long time. The things you do have to be approached the same way they are in a small town. You can't really step on anybody's toes too much because you're going to have to live with them.

The last point is what I call the kill the messenger syndrome. If there is something wrong with the thesis, with the system, with the reagent, the postdoc will be the person who finds out first. Now, depending on who receives that information, either things get changed appropriately, or the kill the messenger syndrome kicks in, which is what happened in the second example I gave. It's something that needs to be avoided, but is very difficult for all humans to resist, regardless of the field. It's much easier to blame the person than blame the process. It is another issue that needs to be addressed. All of you are probably sitting here saying that's really awful, she went to Harvard, we can figure out where this was, this must be really terrible. They should be auditing them like crazy; it doesn't happen in my institution; there's no way we would do this to any of our fellows. I actually think that it is much more widespread than my three or four institution experience. It is at most of the major institutions. The fact is that this often times occurs in more subtle situations. Say, for example, while a postdoctoral fellow is working on a big project, the lab head decides to collaborate with someone else in order to get some reagents. When it comes time for the first and second draft of the publication, she gets told by the head of the lab, we need to add Barney Smith's name onto the authors. The postdoc says: "But why? He didn't do anything but give us some reagents." The lab head replies: "Well, we always do that." In that circumstance, can the postdoc actively resist something that is probably ethically gray, if not definitely against all published recommendations about authorship? Usually those discussions don't go any further than that.

So in talking about all this, you know this may be an exercise in futility. I feel in some ways like I'm bringing ice to Eskimos. If you're here and you're listening, you probably care. Unfortunately, postdoctoral fellows are existing in some suboptimal situations all across the country. I think that even though the selection bias exists and the sample was small and it wasn't a truly random group, the fact that only a couple of us could produce almost 20 names from multiple institutions of people that had serious conflicts and not a single one of them was willing to speak for fear of reprisal, is a telling statement. So why did *I* come forward? Why am *I* here? Am *I* one of those people buried in these stories? Yes, I am. And the reason I'm here is because my boss coerced me into doing it.

Lisa Backus

I would just like to say that I have the same boss that Jan has and he is very persuasive.

When I agreed to speak on this panel, Sigma Xi sent me a summary of the discussion that had happened at the Wingspread conference. In that summary, one of the "major new subjects" for discussion was postdoc problems and under that category they listed interactions between junior and senior investigators. There was a telling sentence in the summary. It said "the fundamental problem is mentoring." Under that summary, it also said that in exit interviews of Ph.D.s, the ones most likely to express dissatisfaction with mentoring were women in life sciences and women in public health. In my brief talk today, I'd like to try to speak to the issue of mentoring, to the issue of gender, and particularly to the area where these two intersect. From my introduction, you know that I completed my Ph.D. I was one of those women in life sciences. Toward the end of my Ph.D., I applied for postdocs and I also applied for admission to medical schools. As my biography also makes clear, I opted to go to medical school. I believe that it's no exaggeration to say that in large part the reason that I ended up at medical school, and not in a postdoc, was because of issues of mentoring and issues of gender. Unfortunately, when I talk with my female friends and colleagues, who instead went on to do postdocs, I have become increasingly convinced that I made the right decision.

Graduate students and postdocs often decry the paucity of good mentors. I see from the schedule that tomorrow there will be an entire panel on the issue of mentoring. I cannot overemphasize the importance of this issue. The question becomes first of all, what is a good mentor, because maybe if we knew what a good mentor was, we could figure out why there are so few of them and then maybe we could figure out why there are especially so few of them for women. In terms of what a good mentor is, as a graduate student you need a mentor in general to teach you to do good science, to come up with ideas, to write papers, and to write grants. Towards the end of the Ph.D., and especially in a postdoc, however, you need a mentor to do something more. It is this intangible that is, I think, the real crux of the problem. The mentor must steer you through the politics of modern science in order to avoid endless postdocs or in order to keep you in science and not leave it altogether. You need a mentor, basically, to look out for your career. He, and I use the pronoun he even though I would much rather use the pronoun she, but looking out over this audience I am reminded that science is a male dominated field. So he, your mentor, needs to make sure that you publish in the right places, that you meet the right people at conferences, that you talk to the right editors at conferences and that when push comes to shove, and you're faced with the prospects that Garth described of looking for a job, that your mentor will actually call people and network for you, see who has jobs, make sure your name gets brought up in conversations. Basically, a good mentor has to get you into the "old boy" network. Carl Djerassi used the term the "tribe," which I contend may be just a more generous or less gender specific version. It is an "old boys" network. It always was. It still is. It is changing, but we should not kid ourselves. It is difficult enough to find a Principal Investigator (P.I.) who will do all of this, networking at conferences and making sure your publications are in the right places, and unfortunately, the issue of gender only makes it more difficult. Two sociologists, Catherine Ward and Linda Grant, are very concerned with the issues of mentoring. They have surveyed over 600 young scientists about mentoring. They found that a significantly greater number of women report a lack of a mentor. Of women who do find mentors, most women find them later in their careers. This lack of mentoring has noticeable effects on the scientific careers of women.

On a personal level, a good friend of mine, now in her second postdoc and facing the prospect of a third postdoc, described her predicament to me recently. She explained that at her lab, her P.I. in general has "gone to bat" for some other postdocs, doing all the things that I described above with networking at conferences, etc. But when she started to look for a more permanent position, it became clear that he wasn't doing that for her and she was baffled. She had done good work. She had a respectable list of publications in respectable journals. It eventually dawned on her that the problem was baseball. She didn't follow baseball. If she couldn't talk about the pitching problems of the Oakland A's with her supervisor, she just wasn't quite one of the gang. She didn't get introduced at conferences as much, she didn't get her name bantered around in conversations quite as much, and it finally became also apparent to her that, in some small way, he felt just ever so slightly uncomfortable around her because he didn't quite know what to talk about with her if he couldn't talk about baseball. Her experience, unfortunately, is born out by general statistics about women in science.

When women started entering graduate schools in large numbers in the early 1970s, most of us believed that it would only be a matter of time before women made their way up the hierarchy. Today, the number of women getting Ph.D.s has grown in almost every field of science. In 1979, 21 percent of new science Ph.D.s went to women. By 1989, that figure was up to 28 percent.

Arthur Singer in his introduction correctly pointed out that in fact women go on in about roughly equal numbers to postdocs as men do. But unfortunately when you start looking at the higher levels, women aren't making it up the ladder. National Science Foundation figures show that in 1979, women held 5 percent of all tenured positions. By 1989, 12 years latter, the figure had risen to only 7 percent. Further analysis by the NSF showed that the sex difference in rank and tenure persists, even when men and women are matched for field of science, for the quality of institution where they earned their doctorate degree, and for the number of years since their doctorate was earned. One of the researchers at the NSF has actually been quite vocal about this situation. She contends that the reason for this difference is very clear. It is because the "old boy" network is very much in place.

Finally, I would like to add that I believe we are at a critical time in terms of mentoring and in terms of women. I fear that things are going to get worse before they get better. In recent months there has been a great deal of publicity around issues of sexual harassment. The Anita Hill/Clarence Thomas hearings, the Tailhook incidents, Francis Conley's public expose of affairs at Stanford have all helped bring the issue of sexual harassment into the public eye. This is, in general, undoubtedly a good thing. The Wingspread Sigma Xi summary noted that sexual harassment was another major problem for postdocs. What I'm concerned about though is what is commonly referred to as a "chilling effect." It is this chilling effect that now might make it even more difficult for women to get the kind of political mentoring that they need. In general, most men feel more comfortable mentoring other men. Concerns about sexual harassment are likely to decrease the casual conversations on issues of baseball and other things outside of science. It is these casual conversations and the bond that they forge between a P.I. and postdoc that ultimately lead the P.I. to lobby on the postdoc's behalf. Thus, concerns about sexual harassment may actually chill these types of casual conversations, only making the mentoring situation worse.

If this were the usual type of scientific communication, I would have a summary slide here. I don't have the slide, but I can tell you what would be on it. There would be three things. One, there is a shortage of mentors for post-docs. Two, the shortage is even more severe for women. And finally three, this shortage has a detrimental effect on the scientific careers for women. I do want to close on an optimistic note. As Jan already pointed out, obviously here I'm preaching to the converted. My hope is that forums such as this will help raise the issues, devise solutions, and then ultimately work for their implementation.

Tara Meyer

First let me begin by saying that I enjoy being a postdoc and that I believe that it is a valuable professional experience. The benefits of post docing are numerous. Not only does a postdoc learn about a different area of science, but he/she will also be exposed, hopefully, to an entirely different way of approaching a problem. A postdoctoral position also gives the academically-bound scientist the opportunity to see how another institution differs from his/her doctoral university. And finally, a postdoctoral advisor typically helps the postdoc find his/her permanent employment and can remain a valuable contact throughout the postdoc's career.

This is sort of the ideal case, however. Not surprisingly, not all postdocs enjoy these benefits. Clearly success/happiness in any job requires a combination of effort, talent and good will from both employer and employee. Where postdoc-ing differs, in my opinion, from a typical job is that the organization of the academic scientific world is feudal. Professors operate research groups as nearly independent enclaves. The departments decide who gets a domain and how big it is, but they do not routinely involve themselves in the day-to-day operations of the group. Groups comprise undergraduates, graduate students, and postdocs. And while all of these types of employees are subject to similar treatment within a particular group, both undergrads and grad students have real status within the university. They are students first, members of a group second and as such have available an extensive university support network. Now this network does not necessarily prevent problems, but it does exist. Postdocs are hired directly by a particular professor without the approval or even knowledge of the department. Their loose association with a particular department ends with the termination of their employment with their advisor — they have no independent status.

Given this system it is necessary that the happiness/success of a postdoc depend almost completely on a complex set of mutual obligations between the postdoc and the advisor. The obligations are clearly two-sided, but I will speak only to those of the postdoctoral advisor. I will limit myself for two reasons. First an advisor who does not fulfill his/her obligations can cripple a postdoc's career whereas one bad postdoc should not significantly harm an established professor. Second, I will not address postdoctoral obligations because, as a postdoc, I do not believe that I could do the subject justice.

1. Discrimination, Sexual harassment, etc. - These behaviors are not permissible in any job situation.

2. The Project - A perspective postdoc should have an accurate picture of what the project will entail, within reason. It is extremely disheartening to a postdoc to discover upon arrival that you have changed the project substantially. The postdoc might not have accepted the position if he/she had known about the change.

3. Support for the project - Sufficient resources should be provided for the project. There is also a time factor involved. A postdoc should not be asked to wait eight months for the arrival of a piece of equipment which is absolutely essential to a project. Clearly, there are always unforeseen delays, but you should have an infrared spectrometer (IR) if you are asking the postdoc to do IR studies.

4. Group duties - The level of group responsibilities varies greatly from group to group. The extent of these duties should be explained clearly and their execution should leave sufficient time so the postdoc can do his/her own research.

5. Project Success - If a project is failing miserably, do not simply assume that the postdoc is at fault. Discuss options with the postdoc for modifying the project such that the probability of producing results increases. Do not force a postdoc to spend two years working on a bad project in order to avoid admitting that the original idea was at fault.

6. Hiring/Termination flexibility - Both incoming and exiting postdocs need flexibility. Neither finishing a thesis, nor getting a job can always be planned accurately. Also if your financial situation precludes flexibility, inform the postdoc as soon as possible.

7. Work Hours - While it is reasonable to expect that a career scientist will be willing to work more than the arbitrary 40 hours per week, it is not reasonable to expect them in every evening and weekend. Discuss work expectations with your postdoc, but be realistic. Also flexibility in work hours is a valuable fringe benefit which can be offered at no cost to you.

8. Be a Mentor - Give encouragement and praise where it is appropriate. Help them get jobs. Teach them to be good researchers. Your students and postdocs are your "descendents" and through them your contributions to science and your reputation will continue long past the end of your career.

9. Project direction - As the postdoc becomes familiar with the project, give them more control. They are Ph.D. chemists and you will probably achieve better results if you collaborate with them rather than using them as technicians.

10. Acknowledgement of Work - Have a clear policy about how contributions are acknowledged and discuss it with your postdocs. If a postdoc contributes a truly original idea or suggests an original project, consider allowing them to publish independently from you. Do not simply adopt their ideas and write them into future grant proposals and papers.

Clearly, due to the organization of academic research groups, these guidelines can be put forth only as suggestions. Granting agencies can put forth certain employment rules, but they do not cover many of the more complex issues such as mentoring. I do, however, believe that individual departments could increase their involvement in these areas by, perhaps, issuing a similar set of guidelines and offering mediation/counseling to postdocs and supervisors who are having problems.

In summary, the postdoctoral position in chemistry is unique in that the postdoc is employed by a single individual whose employment practices are largely unmonitored. I have enumerated several of what I believe are the postdoctoral supervisor's obligations which although largely "understood" in the academic world are not universally practiced and are rarely if ever discussed. I also believe that increased interest by academic departments in these issues could provide some much needed standardization of postdoctoral working conditions.

Scientific Elites and Scientific Illiterates

David L. Goodstein

Scientific papers often begin by posing a paradox, even if it is one that had not previously seemed particularly disturbing. Having posed the paradox, the author then proceeds to resolve it. At first glance, we don't seem to make much progress that way. A paradox that was previously unnoticed is now no longer unexplained. However, such exercises can sometimes be useful. For example, Albert Einstein's famous 1905 paper introducing the theory of relativity was very much of this form. He began by pointing out that when a magnet induces an electric current in a loop of wire, we attribute that effect to entirely different causes depending on whether the magnet or the loop of wire is in motion. Finding this paradox intolerable, he proceeded to resolve it, giving new meanings to time and space along the way.

Today, with my customary modesty, I would like to follow in Albert's bicycle tracks (see Figure 1) and begin by posing a paradox. The paradox is that we, here in the United States today, have the finest scientists in the world, and we also have the worst science education in the world, or at least in the industrialized world. There seems to be little doubt that both of these seemingly contradictory observations are true. American scientists, trained in American graduate schools produced more Nobel Prizes, more scientific citations, more of just about anything you care to measure than any other country in the world; maybe more than the rest of the world combined. Yet, students in American schools consistently rank at the bottom of all those from advanced nations in tests of scientific knowledge, and furthermore roughly 95 percent of the American public is consistently found to be scientifically illiterate by any rational standard. How can we possibly have arrived at such a result? How can our miserable system of education have produced such a brilliant community of scientists? I would like to refer to this situation as The Paradox of the Scientific Elites and the Scientific Illiterates.

In my view, these two seemingly contradictory observations are both true, and they are closely related to one another. We have created a kind of feudal aristocracy in American science, where a privileged few hold court, while the toiling masses huddle in darkness, metaphorically speaking, of

Figure 1

course. However, I also think inexorable historic forces are at work that have already begun to bring those conditions to an end. Not that light will be brought to the masses necessarily, but that our days at court are clearly numbered. To understand all this, and before we get any more deeply mired in dubious metaphors, it may help to go back to the beginning. I mean, really The Beginning.

In modern Cosmology, the accepted theory of the beginning of the universe goes something like this: At a certain instant around 10 to 15 billion years ago, the universe was created in a cataclysmic event called The Big Bang. It has been expanding uniformly ever since. What we do not know, however, is whether the density of matter in the universe is great enough to reverse that expansion eventually, causing the universe to slow down, come to a stop, and then finally fall back upon itself. If that does happen, the cosmologists are prepared with a name for the final cataclysmic moment when the universe ends. It will be known as The Big Crunch.

Today I would like to offer you a somewhat analogous theory of the history of science. According to this theory, science began in a cataclysmic event sometime around the year 1700 (the publication of Newton's *Principia* in 1687 is a good candidate for the actual event). It then proceeded to expand at a smooth, continuous exponential rate for nearly 300 years. Unlike the universe, however, science did not expand into nothing at all. Instead, the expansion must come to an end when science reaches the natural limits imposed on it by the system it was born into, which is called The Human Race.

I don't mean that scientific knowledge is limited by the human race; in fact, I don't think scientific knowledge is limited at all, and I hope that will go on expanding forever. What I'm talking about here is what you might call the profession of science, or the business of science. It is my opinion that the size of the scientific enterprise, which began its exponential expansion around 1700, has now begun to reach the limits imposed on it by the size of the human race. Thus, the expansion of science is now in the process of ending, not in a Big Crunch, but in something much more like a whimper, that may or may not leave some residue of science still existing when it is all over. I think that the beginning of the end of the exponential expansion era of science occurred, in the United States at least, around the year 1970. Most people, scientists and otherwise, are unaware that it is coming to an end (in fact, they probably never knew it existed) and are still trying to maintain a social structure of science (by which I mean research, education, funding, institutions and so on) that is based on the unexamined assumption that the future will be just like the past. Since I believe that to be impossible, we have some very interesting times ahead of us. I would like to tell you today why I believe all this, and what we might try to do about it.

Figure 2 is borrowed from a book called *Little Science, Big Science* by Derek da Solla Price. Price may be identified as the Edwin Hubble of the expansion of science (Hubble discovered the expansion of the Universe). The

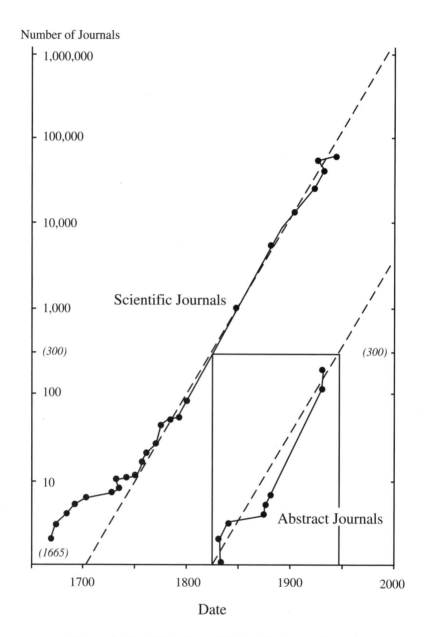

***Figure 2: Total Number of Scientific Journals and
Abstract Journals Founded, as a Function of Date***

From *Little Science, Big Science* by Derek da Solla Price. Columbia University Press, 1986, p. 8.

figure, a plot of the number of scientific journals founded, world wide, as a function of year, is a suitable stand-in for any other quantitative measure of the size of science. It shows that the cumulative number of journals founded increased by a factor of 10 about every 50 years, from 1760 to 1950. This is a different, faster kind of growth than a free expansion like that of the universe. Here the rate of growth of the system keeps increasing as the size of the system increases. In other words, the bigger it is, the faster it grows. Anyone observing this so-called exponential curve would conclude that science was born (roughly) in the year 1700, and that a million journals would have been founded by the year 2000. Price, who pointed out this phenomenon in the early 1960's, was clever enough to know that neither of these conclusions would be correct. On the one hand, both scientific knowledge and the scientific enterprise have roots that stretch all the way back to antiquity, and on the other hand the number of scientific journals in the world today, as we approach the year 2000, is a mere 40,000. This sorry failure of the publishing industry to keep up with our expectations often leaves us scientists with nothing to read by the time we reach the end of the week.

The point is that the era of exponential growth in science is already over. The number of journals is one measure, but all others tend to agree. In particular, it applies to the number of scientists around. It is probably still true that 90 percent of all the scientists who have ever lived are alive today, and that statement has been true at any given time for nearly 300 years. But it cannot go on being true for very much longer. Even with the huge increase in world population in this century, only about one-twentieth of all the people who have ever lived are alive today. It is a simple mathematical fact that if scientists keep multiplying faster than people, there will soon be more scientists than there are people. That seems very unlikely to happen.

I think the last 40 years, in the United States, have seen the end of the long era of exponential growth and the beginning of a new era we have not yet begun to imagine. These years will be seen in the future as the period in which science began a dramatic and irreversible change into an entirely new regime. Let's look back at what has happened in those 40 years in light of this historic transformation.

The period 1950-1970 was a true golden age for American science. Young Ph.D.'s could choose among excellent jobs, and anyone with a decent scientific idea could be sure of getting funds to pursue it. The impressive successes of scientific projects during the Second World War had paved the way for the federal government to assume responsibility for the support of basic research. Moreover, much of the rest of the world was still crippled by the after-effects of the war. At the same time, the G.I. Bill of Rights sent a whole generation back to college. The American academic enterprise grew explosively, especially in science and technology. Even so, that explosive growth was merely a seamless continuation of the exponential growth of science that had dated back to 1700. It seemed to one and all (with the notable exception of Derek da Solla Price) that these happy conditions would go on forever.

By now, in the 1990's, the situation has changed dramatically. With the Cold War over, national security is rapidly losing its appeal as a means of generating support for scientific research. To make matters worse, the country is 4 trillion dollars in debt and scientific research is among the few items of discretionary spending in the national budget. There is much wringing of hands about impending shortages of trained scientific talent to ensure the nation's future competitiveness, especially since by now other countries have been restored to economic and scientific vigor, but in fact, jobs are scarce for recent graduates. The best American students have proved their superior abilities by reading the handwriting on the wall and going into other lines of work. Half the students in American graduate schools in science and technology are from abroad. The golden age definitely seems over.

Both periods, the euphoric golden age, 1950-1970, and the beginning of the crunch, 1970-1990, seemed at the time to be the product of specific temporary conditions rather than grand historic trends. In the earlier period, the prestige of science after helping win the war created a money pipeline from Washington into the great research universities. At the same time, the G.I. Bill of Rights transformed the United States from a nation of elite higher education to a nation of mass higher education. Before the war, about 8 percent of Americans went to college, a figure comparable to that in France or England. By now more than half of all Americans receive some sort of post-secondary education, and nearly a third will eventually graduate from college. To be sure, this great and noble experiment in mass higher education has failed utterly and completely in technology and science, where 4-5 percent of the population can be identified as science and technology professionals, and the rest may as well live in the pre-Newtonian era. Nevertheless, the expanding academic world in 1950-1970 created posts for the exploding number of new science Ph.D.'s, whose research led to the founding of journals, to the acquisition of prizes and awards, and to increases in every other measure of the size and quality of science. At the same time, great American corporations such as AT&T, IBM and others decided they needed to create or expand their central research laboratories to solve technological problems, and also to pursue basic research that would provide ideas for future developments. And the federal government itself established a network of excellent national laboratories that also became the source of jobs and opportunities for aspiring scientists. As we have already seen, all this extraordinary activity merely resulted in a 20-year extension in the U.S. of the exponential growth that had been quietly going on since 1700. However, it was to be the last 20 years. The expansionary era in the history of science was about to come to an end, at least in America.

Actually, during the second period, 1970-1990, the expansion of American science did not stop altogether, but it did slow down significantly compared to what might have been expected from Price's exponential curves. Federal funding of scientific research, in inflation-corrected dollars, doubled during that period, and by no coincidence at all, the number of academic researchers also doubled. Such a controlled rate of growth (controlled only by the available funding, to be sure) was not, however consistent with the lifestyle that academic researchers had evolved. The average American professor in a research university turns out about 15 Ph.D. students in the course of a career. In a stable, steady-state world of science, only one of those 15 can go on to become another professor in a research university. In a steady-state world, it is mathematically obvious that the professor's only reproductive role is to produce one professor for the next generation. But the American Ph.D. is basically training to become a research professor. American students, realizing that graduate school had become a training ground for a profession that no longer offered much opportunity, started choosing other options. The impact of this situation was obscured somewhat by the growth of postdoctoral research positions, a kind of holding tank for scientific talent that allowed young researchers to delay confronting reality for three or six or more years. Nevertheless, it is true that the number of the best American students who decided to go to graduate school started to decline around 1970, and it has been declining ever since.

In the meantime, yet one more surprising phenomenon has taken place. The golden age of American academic science produced genuine excellence in American universities. Without any doubt at all, we lead the world in scientific training and research. It became necessary for serious young scientists from everywhere else either to obtain an American Ph.D., or at least to spend a year or more of postgraduate study here. America has come to play the role for the rest of the world, especially the emerging nations of the Pacific rim, that Europe once played for young American scientists, and it is said, that Greece once played for Rome. We have become the primary source of scientific culture and learning for everyone. Almost unnoticed, over the past 20 years the missing American graduate students have been replaced by foreign students. This change has permitted the American research universities to go on producing Ph.D.'s almost as before.

Nevertheless, it should be clear by now that with half the kids in America already going to college, academic expansion is finished. With the Cold War over, competition in science can no longer be sold as a matter of national survival. There are those who argue that research is essential for our economic future, but the managers of the economy know better. The great corporations have decided that central research laboratories were not such a good idea after all. Many of the national laboratories have lost their missions and have not found new ones. The economy has gradually transformed from manufacturing to service, and service industries like banking and insurance don't support much scientific research.

Each of these conditions appears to be transient and temporary, but they are really the immediate symptoms of a large-scale historic transformation. For us in the United States, the expansionary era of the history of science has come to an end. The future of American science will be very different from the past.

Let's get back now to the Paradox of Scientific Elites and Scientific Illiterates. The question of how we educate our young in science lies at the heart of the issues we have been discussing. The observation that, for hundreds of years

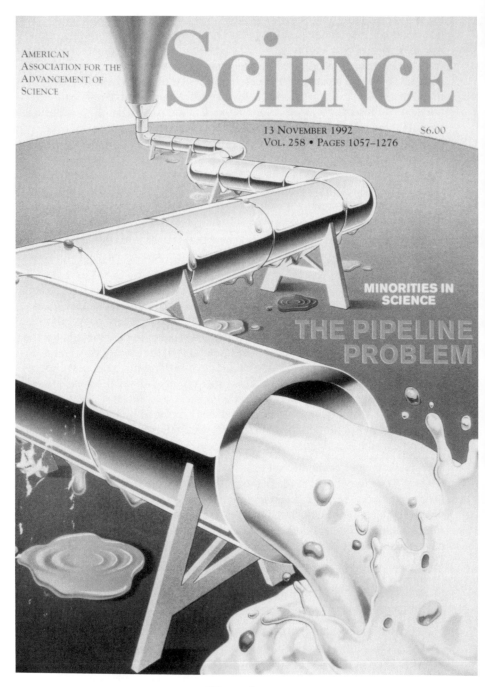

Figure 3

the number of scientists had been growing exponentially means, quite simply, that the rate at which we produced scientists has always been proportional to the number of scientists that already existed. We have already seen how that process works at the final stage of education, where each professor in a research university turns out 15 Ph.D.'s, most of those wanting to become research professors and turn out 15 more Ph.D.'s.

Recently, however, a vastly different picture of science education has been put forth and has come to be widely accepted. It is the metaphor of the pipeline, illustrated in Figure 3, which shows the cover of a recent issue of *Science* magazine. The idea is that our young people start out as a torrent of eager, curious minds anxious to learn about the world, but as they pass through the various grades of schooling, that eagerness and curiosity is somehow squandered, fewer and fewer of them showing any interest in science, until at the end of the line, nothing is left but a mere trickle of Ph.D.'s. Thus, our entire system of education is seen to be a leaky pipeline, badly in need of repair. However, the artist who drew the *Science* cover doesn't seem to have grasped the idea that only a trickle should emerge at the end. As the cover of *Science* indicates the leakage problem is seen as particularly severe with regard to women and minorities, but the pipeline metaphor applies to all. I'm not quite sure, but I think the pipeline metaphor came first out of the National Science Foundation, which keeps careful track of science workforce statistics (at least that's where I first heard it). As the NSF points out with particular urgency (and the *Science* cover echoes) women and minorities will make up the majority of our working people in future years. If we don't figure out a way to keep them in the pipeline, where will our future scientists come from?

I believe it is a serious mistake to think of our system of education as a pipeline leading to Ph.D.'s in science or in anything else. For one thing, if it were a leaky pipeline, and it could be repaired, then as we've already seen, we would soon have a flood of Ph.D.'s that we wouldn't know what to do with. For another thing, producing Ph.D.'s is simply not the purpose of our system of education. Its purpose instead is to produce citizens capable of operating a Jeffersonian democracy, and also if possible, of contributing to their own and to the collective economic well being. To regard anyone who has achieved those purposes as having leaked out of the pipeline is worse than arrogant; it is silly. Finally, the picture doesn't work in the sense of a scientific model: it doesn't make the right predictions. We have already seen that, in the absence of external constraints, the size of science grows exponentially. A pipeline, leaky or otherwise, would not have that result. It would only produce scientists in proportion to the flow of entering students.

I would like to propose a different and more illuminating metaphor for science education. It is more like a mining and sorting operation, designed to cast aside most of the mass of common human debris, but at the same time to discover and rescue diamonds in the rough, that are capable of being cleaned and cut and polished into glittering gems, just like us, the existing scientists. It takes only a little reflection to see how much more this model accounts for than the pipeline does. It accounts for exponential growth, since it takes scientists to identify prospective scientists. It accounts for the very real problem that women and minorities are woefully underrepresented among scientists, because it is hard for us, white, male scientists to perceive that once they are cleaned, cut and polished they will look like us. It accounts for the fact that science education is for the most part a dreary business, a burden to student and teacher alike at all levels of American education, until the magic moment when a teacher recognizes a potential peer, at which point it becomes exhilarating and successful. Above all, it resolves the paradox of Scientific Elites and Scientific Illiterates. It explains why we have the best scientists and the most poorly educated students in the world. It is because our entire system of education is designed to produce precisely that result.

It is easy to see the sorting operation at work in the college physics classroom, where most of my own experience is centered, but I believe it works at all levels of education and in many other subjects. From elementary school to graduate school, from art and literature to chemistry and physics, students and teachers with similar inclinations resonate with one another. The tendency is natural and universal. But, if it is so universal, you might ask, why is America so much worse off than the rest of the world? The answer, I think, is that in education and in science, as in fast food and popular culture, America is not really worse than the rest of the world, we are merely a few years ahead of the rest of the world. What we are seeing here will happen everywhere soon enough. Our colleagues abroad can take what scant comfort they can find in the promise that our dilemmas in science and education are on the way, along with Big Macs and designer jeans.

Getting back to America, the mining and sorting operation that we call science education begins in elementary school. Most elementary school teachers are poorly prepared to present even the simplest lessons in scientific or mathematical subjects. In many places, Elementary Education is the only college major that does not require even a single science course, and it is said that many students who choose that major do so precisely to avoid having to take a course in science. To the extent that is true, elementary school teachers are not merely ignorant of science, they are preselected for their hostility to science, and no doubt they transmit that hostility to their pupils, especially young girls for whom elementary school teachers must be powerful role models. Even those teachers who did have at least some science in college are not likely to be well prepared to teach the subject. Recently, I served on a kind of visiting committee for one of the elite campuses of The University of California, where every student is required to have at least one science course. The job of the committee was to determine how well this requirement was working. We discovered that 90 percent of the students in majors outside science and technology were satisfying the requirement by taking a very popular biology course known informally as "human sexuality." I don't doubt for an instant that the course was valuable and interesting, and may even have tempted the students to do voluntary "hands on" experimentation on their own time (a result we seldom achieve in physics). But I do not think that such a course by itself offers sufficient training in science for a university graduate at the end of the 20th century. These students, some of whom will go on to become educators, are themselves among the discards of the science mining and sorting operation.

In any case, the first step of the operation is what might be called passive sorting, since few elementary school pupils come into personal contact with anyone who has scientific training. Certainly, we all know that many young people decide that science is beyond their understanding long before they have any way of knowing what science is about. Nevertheless, a relatively small number of students, usually those who sense instinctively that they have unusual technical or mathematical aptitudes, arrive at the next levels of education with their interest in science still intact.

The selection process becomes more active in high school. There are about 22,000 high schools in the United States, most of which offer at least one course in physics. Physics is my own subject, and I have had some influence on the teaching of physics in American high schools because a remarkably large fraction of them use "The Mechanical Universe," a television teaching project I directed some years ago. Because I have some first-hand knowledge about physics in high schools, I'll stick to that, although I suspect what I have to say applies to other science subjects as well. Anyway, there are just a few thousand trained high school physics teachers in the U.S., far fewer than there are high schools. The majority of courses are taught by people, who, in college, majored in chemistry, biology, mathematics, or surprisingly often, home economics, a subject that has lost favor in recent years. I know from personal contact that these are marvelous people, often willing to work extraordinarily hard to make themselves better teachers of a subject they never chose for themselves. My greatest satisfaction from making "The Mechanical Universe" comes from the very substantial number of them who have told me that I helped make their careers successful. Their greatest satisfaction comes from — guess what — discovering those diamonds in the rough that can be sent on to college for cutting and polishing into real physicists.

I don't think I need to explain to you what happens in college and graduate school, but I'd like to tell you a story of my own because I think it helps to illustrate one of my main points. By far the best course I had in college was not in physics, but rather it was a required writing and literature course known as Freshman English. The professor was my hero, and I was utterly devoted to him. He responded just as you might expect: he tried hard to talk me into quitting science and majoring in English. Nevertheless, the thought of actually doing that never crossed my mind. I knew perfectly well that if I was ever going to make anything of myself, I was going to have to suffer a lot more than I was doing in Freshman English! The story illustrates that we scientists are not the only ones who engage in mining and sorting. The real point, however, is that for most of us in the academic profession, our real job is not education at all; it is vocational training. We are not really satisfied with our handiwork unless it produces professional colleagues. That is one of the characteristics that may have to change in the coming brave new world of post-expansion science.

American education is much-maligned, and of course it suffers from severe problems that I need not go into here. Nevertheless, it was remarkably well suited to the exponential expansion era of science. Mass higher education, essentially an American invention, means that we educate nearly everyone, rather poorly. The alternative system, gradually going out of style in Europe these days, is to educate a select few rather well. But we too have rescued elitism from the jaws of democracy, in our superior graduate schools. Our students finally catch up with their European counterparts in about the second year of graduate school (this is true, at least, in physics) after which they are second to none. When, after about 1970, the gleaming gems produced by this assembly line at the end of the mining and sorting operation were no longer in such great demand at home, the humming machinery kept right on going, fed by ore imported from across the oceans.

To those of us who are professors in research universities, those foreign graduate students have, temporarily at least, rescued our way of life. In fact we are justly proud that in spite of the abysmal state of American education in general, our graduate schools are a beacon unto the nations of the world. The students who come to join us in our research are every bit as bright and eager as the home-grown types they have partially replaced, and they add energy and new ideas to our work. However, there is another way of looking at all this. Graduate students in the sciences are often awarded teaching assistantships, for which they may not be well qualified, because their English is imperfect. In general, through teaching or research assistantships or fellowships, they are paid stipends and their tuitions are either waived, or subsidized by the universities. Thus our national and state governments find themselves supporting expensive research universities that often serve undergraduates poorly (partly because of those foreign teaching assistants) and whose principal educational function at the graduate level has become to train Ph.D.'s from abroad. Some of these, when they graduate, stay on in America, taking some of those few jobs still available here, and others return to their homelands taking our knowledge and technology with them to our present and future economic competitors. It doesn't take a genius to realize that our state and federal governments are not going to go on forever supporting this playground we professors have created for ourselves.

To most of us professors, of course, science no longer seems like a playground. Recently, Leon Lederman, one of the leaders of American science published a pamphlet called *Science — The End of the Frontier*. The title is a play on *Science — The Endless Frontier*, the title of the 1940's report by Vannevar Bush that led to the creation of the National Science Foundation and helped launch the Golden Age described above. Lederman's point is that American science is being stifled by the failure of the government to put enough money into it. I confess to being the anonymous Caltech professor quoted in one of Lederman's sidebars to the effect that my main responsibility is no longer to do science, but rather it is to feed my graduate students' children. Lederman's appeal was not well-received in Congress, where it was pointed out that financial support for science is not an entitlement program, nor in the press, where the *Washington Post* had fun speculating about hungry children haunting the halls of Caltech. Nevertheless, the problem Lederman wrote about is very real and very painful to those of us who find that our time, attention and energy are now consumed by raising funds rather than doing research. However, although Lederman would certainly disagree with me, I firmly believe that this problem cannot be solved by more government money. If federal support for basic research were to be doubled (as many are calling for), the result would merely be to tack on a few more years of exponential expansion before we'd find ourselves in exactly the same situation again. Lederman has performed a valuable service in promoting public debate of an issue that has worried me for a long time (the remark he quoted is one I made in 1979), but the issue itself is really just a symptom of the larger fact that the era of exponential expansion has come to an end.

The crises that face science are not limited to jobs and research funds. Those are bad enough, but they are just the beginning. Under stress from those problems, other parts of the scientific enterprise have started showing signs of distress. One of the most essential is the matter of honesty and ethical behavior among scientists.

The public and the scientific community have both been shocked in recent years by an increasing number of cases of fraud committed by scientists. There is little doubt that the perpetrators in these cases felt themselves under intense pressure to compete for scarce resources, even by cheating if necessary. As the pressure increases, this kind of dishonesty is almost sure to become more common.

Other kinds of dishonesty will also become more common. For example, peer review, one of the crucial pillars of the whole edifice, is in critical danger. Peer review is used by scientific journals to decide what papers to publish, and by granting agencies such as the National Science Foundation to decide what research to support. Journals in most cases, and agencies in some cases, operate by sending manuscripts or research proposals to referees who are recognized experts on the scientific issues in question, and whose identity will not be revealed to the authors of the papers or proposals. Obviously, good decisions on what research should be supported and what results should be published are crucial to the proper functioning of science.

Peer review is usually quite a good way of identifying valid science. Of course, a referee will occasionally fail to appreciate a truly visionary or revolutionary idea, but by and large, peer review works pretty well so long as scientific validity is the only issue at stake. However, it is not at all suited to arbitrate an intense competition for research funds or for editorial space in prestigious journals. There are many reasons for this, not the least being the fact that the referees have an obvious conflict of interest, since they are themselves competitors for the same resources. It would take impossibly high ethical standards for referees to avoid taking advantage of their privileged anonymity to advance their own interests, but as time goes on, more and more referees have their ethical standards eroded as a consequence of having themselves been victimized by unfair reviews when they were authors. Peer review is thus one among many examples of practices that were well-suited to the time of exponential expansion, but will become increasingly dysfunctional in the difficult future we face.

We must find a radically different social structure to organize research and education in science. That is not meant to be an exhortation. It is meant simply to be a statement of a fact known to be true with mathematical certainty, if science is to survive at all. The new structure will come about by evolution rather than design, because, for one thing, neither I nor anyone else has the faintest idea of what it will turn out to be, and for another, even if we did know where we are going to end up, we scientists have never been very good at guiding our own destiny. Only this much is sure: the era of exponential expansion will be replaced by an era of constraint. Because it will be unplanned, the transition is likely to be messy and painful for the participants. In fact, as we have seen, it already is. Ignoring the pain for the moment, however, I would like to look ahead and speculate on some conditions that must be met if science is to have a future as well as a past.

It seems to me that there are two essential and clearly linked conditions to consider. One is that there must be a broad political consensus that pure research in basic science is a common good that must be supported from the public purse. The second is that the mining and sorting operation I've described must be discarded and replaced by genuine education in science, not just for the scientific elite, but for all the citizens who must form that broad political consensus.

Basic research is a common good for two reasons: it helps to satisfy the human need to understand the universe we inhabit, and it makes new technologies possible. It must be supported from the public purse because it does not yield profits if it is supported privately. Because basic research in science flourishes only when it is fully open to the normal processes of scientific debate and challenge, the results are available to all. That is why it is always more profitable to use someone else's basic research than to support your own. For most people it will also always be easier to let someone else do the research. In other words, not everyone wants to be a scientist. But to fulfill the role of satisfying human curiosity, which means something more than just our own, we scientists must find a way to teach science to non-scientists.

That job may turn out to be impossible. Perhaps professional training is the only possible way to teach science. There was a time long ago when self-taught amateurs could not only make a real contribution to science, but could even become great scientists. Benjamin Franklin and Michael Faraday come to mind immediately. That day is long gone. I get manuscripts in the mail every week (attracted, no doubt, by my fame as a T.V. star) from amateurs who have made some great discovery that they want me to bring to the attention of the scientific world, but they are always nonsense. The frontiers of science have moved far from the experience of ordinary persons. Unfortunately, we have never developed a way to bring people along as informed tourists of the vast terrain we have conquered, without training them to become professional explorers. If it turns out to be impossible to do that, the people may decide that the technological trinkets we send back from the frontier are not enough to justify supporting the cost of the expedition. If that happens, science will not merely stop expanding, it will die.

Tackling in a serious way the as yet remote task of bringing real education in science to all American students would have at least one enormous advantage: it would give a lot of scientists something worthwhile to do. On the other hand, I'm not so sure that opening our territories to tourism will bring unmixed blessings down upon us. For example, would the scientifically knowledgeable citizens of our Jeffersonian republic think it worth $10 billion of public funds to find out what quarks are made of? I don't know the answer to that question, but I am reasonably sure that a scientifically literate public would not have supported President Reagan's Star Wars program, which in its turn, did help for a while to support at least a small part of my own research. In other words, keeping the tourists away has some advantages that we may have to give up.

Nevertheless, I'm willing to take the gamble if you are. I don't think education is the solution to all our problems, but it does seem like a good place to start.

Besides, I really don't know what else we can do.

The Responsibility of Scientists for Science Policy

Thomas F. Malone

As the 20th century draws to a close and we approach the threshold of the third millennium, an increasingly interdependent world is challenged by an intertwined set of attractive opportunities and formidable problems. The roots of these problems and opportunities are embedded deeply in the thought expressed three decades ago by Rachel Carson in her seminal work *Silent Spring* (1). She remarked that "It is only in the moment of time represented by the present century that one species, man, has acquired the power to alter the nature of the world."

The driving forces of this empowerment are found in the progress in developing the four dimensions of knowledge: generation, integration, dissemination, and application (2). Generation is the extension of the frontiers of our understanding of the world we inhabit and our place in that world. It involves basic research in the disciplines that make up the body of knowledge found in the physical, biological, social and engineering sciences, mathematics, and the humanities. Integration is the synthesis of the understanding in these disciplines. It addresses the synergistic interactions among them. Its growing importance is indicated by the emergence of new subdisciplines (e.g., biochemistry) that combine two or more disciplines, and by the increasing emphasis on interdisciplinary studies (3). Dissemination embraces formal education at all levels, and the transfer of understanding outside the classroom. Application involves marshalling the generation, integration, and dissemination of our understanding to meet essential human *needs* (food, clothing, shelter, health, etc.), vital human *aspirations* (education, culture, raison d'être for being, etc.), and human *wants* (e.g., creature comforts, luxuries).

The spectacular and continuing progress made in the natural sciences during the twentieth century suggests that the 21st century *could* be the first age since the dawn of civilization when the total body of human knowledge in all of its four dimensions brings within reach a new vision of world society. It is one that would be a distinct improvement over the one we now have (4).

It is in this context that the responsibility of the scientist for science policy comes into sharp focus. Moreover, it is at this point that the inextricable linkage among science, science policy, and public policy assumes importance. There is, in fact, a continuum embraced by scientific responsibility, science policy, and public policy. Within that continuum there is a hierarchy of ethical issues, ranging from *ordinary*, through *extraordinary*, and on to *very extraordinary*. In the framework of this forum, it is appropriate that we address one of the *very extraordinary* issues that is central to the human prospect in the 21st century.

This issue follows directly from the way in which the "nature of the world" has been altered by the empowerment provided by our still tentative exploration of the domain of knowledge. Progress in the four dimensions of knowledge has been sufficient to enable us to multiply many times the individual capacity to convert the natural resources of the environment into the goods and services that meet essential human *needs*, *aspirations*, and *wants*. This progress has also enabled us to prolong life expectancy at all ages.

The result has been an exponential growth both in the global economy that produces goods and services and in the number of people in the world consuming those goods and using those services. The world population and the world economy have both been doubling every few decades. These rates of growth from the present demographic and economic levels are perturbing the interaction of physical, chemical, and biological processes on planet Earth. The effect is to increase the pressure on the resilient, but not infinite, renewable resources found in the environmental realms of air, water, soil, and plant and animal life.

These pressures are not uniformly distributed because the driving demographic and economic forces have wide geographical variation. In some parts of the world the demographic impact is dominant. In those regions a more-or-less unsuccessful struggle to satisfy essential human needs of a large and rapidly growing population is degrading the productive capacity of the environment and jeopardizing its capacity to support present and future generations. Soil erosion, deforestation, desertification and loss of biodiversity are critical and increasing problems. Over one-half of the world's population lives in these regions. Nineteen out of every 20 individuals added to the world population are born there. Poverty is endemic. The average value of goods and services produced and consumed daily by each person is less than one dollar. More than one billion individuals are ill-fed, ill-clothed, ill-housed, uneducated, and without proper medical care (5).

In other regions, the impact is mainly the result of a large and rapidly expanding economy driven by a combination of human needs, aspirations and seemingly insatiable wants, and powered by advances in the utilization of energy and technology. Although less than 18 percent of the world's population reside in these regions, two-thirds of the world's production and consumption of goods and services takes place there. A powerful, growing, energy- and technology-driven economy is beginning to stress the productive and assimilative capacity of the global environment. The reality of depletion in stratospheric ozone and the threat of global greenhouse-gas warming are manifestations of these stresses. The average value of the daily production and consumption of goods and services by each person is about $30 (in 1980 dollars). Of course, extremes of affluence and poverty exist in these regions, just as they do in those regions where the demographic influence is dominant.

Within this overall context, the likely prospects for the 21st century are shown in the scenarios in Figures 1-4. Some modification of the trajectory on which society is presently embarked has been incorporated into these scenarios. Slower population growth and accelerated economic development have been provided in regions where these changes appear to be desirable. Moderation in the rate of economic growth has been incorporated into the scenarios for regions where that factor is beginning to present problems.

These scenarios have been adapted from data assembled by the World Bank (6) and analysis at the International Institute of Applied System Analysis in Vienna of options for countering greenhouse-gas warming (7). They are presented here not as another "doomsday scenario" currently so popular, but as tools of potential value in assessing the responsibility of the scientist to reflect on his or her role in helping to shape those aspects of science policy that are germane to public policy, with profound implications for the human prospect.

Figure 1 compares actual population in 1990 with a likely scenario for 2050 in (i) the world at two different rates of growth, (ii) 25 high-income countries, (iii) 60 middle-income countries, (iv) 40 low-income countries, and (v) China and three central-planning countries (North Korea, Mongolia, and Vietnam) as a special case of low-income countries. World population would increase threefold if the present annual growth rate (1.7 percent) continues. Even with a reduction in fertility rates, population would double by 2050. The increase in the population in the low-income countries between 1990 and 2050 would be three times the total population in the high-income countries in 2050. This growth would be about equal to the total world population in 1960. This scenario would present serious problems for the creation of the socio-economic infrastructure required to support such an augmentation of population in a region now plagued by poverty.

Figure 2 depicts the likely economic changes for the world and for the several regions if growth is accelerated in the low- and middle-income countries and moderated in the high-income countries. Worldwide production and consumption would grow fourfold in this scenario. If the global economy were to grow at the traditional target of three percent annually, it would increase sixfold by 2050. The *increase* in the conversion of natural resources into goods and services in the high-income countries between 1990 and 2050 would be seven times the *level* of this activity by 2050 in the low-income countries. A simple calculation yields the result that the world economy would have to grow by about 18 times if the level of economic activity in all regions were to equal that in the high-income countries. In that case, the global value of the

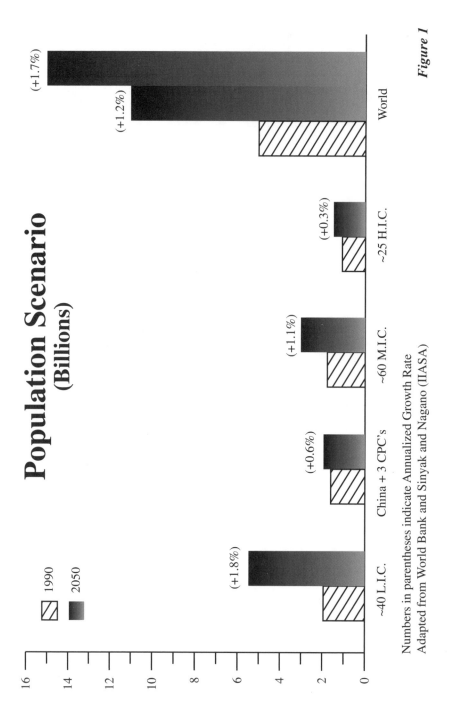

Figure 1

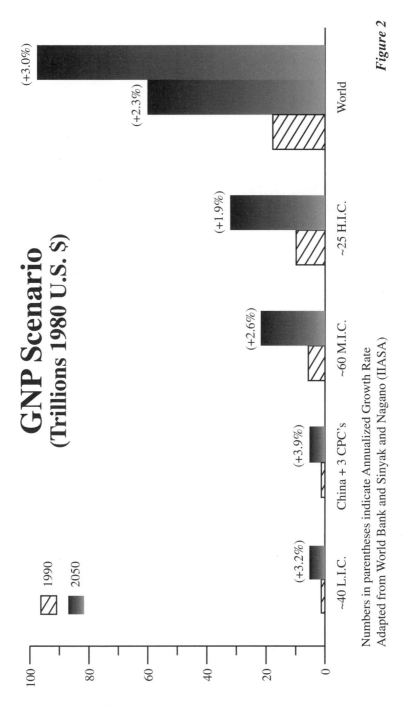

GNP Scenario
(Trillions 1980 U.S. $)

1990
2050

~40 L.I.C.   (+3.2%)

China + 3 CPC's   (+3.9%)

~60 M.I.C.   (+2.6%)

~25 H.I.C.   (+1.9%)

World   (+3.0%)   (+2.3%)

Numbers in parentheses indicate Annualized Growth Rate
Adapted from World Bank and Sinyak and Nagano (IIASA)

Figure 2

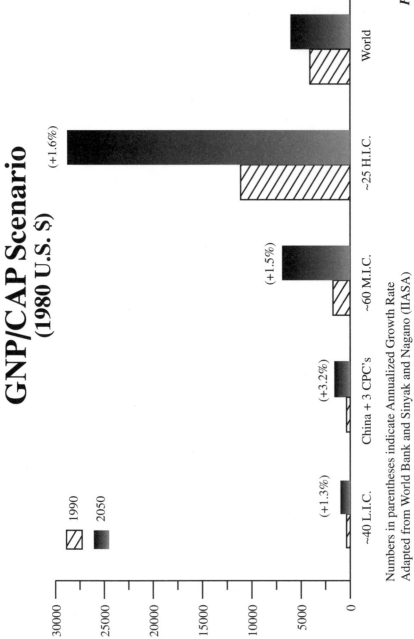

GNP/CAP Scenario
(1980 U.S. $)

Numbers in parentheses indicate Annualized Growth Rate
Adapted from World Bank and Sinyak and Nagano (IIASA)

Figure 3

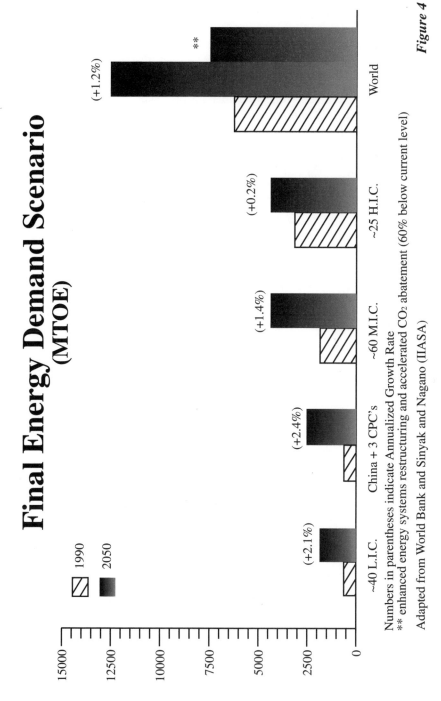

Final Energy Demand Scenario
(MTOE)

1990

2050

~40 L.I.C.   China + 3 CPC's   ~60 M.I.C.   ~25 H.I.C.   World

(+2.1%)   (+2.4%)   (+1.4%)   (+0.2%)   (+1.2%)

Numbers in parentheses indicate Annualized Growth Rate
** enhanced energy systems restructuring and accelerated CO_2 abatement (60% below current level)

Adapted from World Bank and Sinyak and Nagano (IIASA)

Figure 4

production and consumption of goods and services would grow from $16 trillion in 1990 to $286 trillion by 2050 (measured in 1980 U.S. dollars).

The combined effect of asymmetrical demographic and economic growth is shown in Figure 3 which portrays the economic levels of gross national product per person (GNP/CAP). The incremental economic growth per person in the 25 high-income countries would be 50 times that in the 40 low-income countries!

A likely scenario for energy demand is presented in Figure 4. With increased energy efficiency, world demand would double as the global economic output quadruples. By reference to Figure 1, it is apparent that the energy demand per person in the 25 high-income countries would be about nine times that in the 40 low-income countries. If the emission of carbon dioxide were to be reduced to the level suggested by the Intergovernmental Panel on Climate Change (8) in order to preclude global warming, there would be only a modest increase in energy demand. This would require major changes in the modes of energy generation and major problems in the manner that these reductions would have to take place in the several regions.

These scenarios present several challenges to an increasingly interdependent world, and therefore to the scientist whose work plays an increasingly important role in human affairs in that world.

First, there is the likely prospect that even with moderated rates of asymmetrical exponential demographic and economic growth, the quality of life would be diminished by escalating environmental degradation. That quality is not all that it could be at the present time, given the advances in the several dimensions of knowledge that we have already achieved, however modest they have been.

Second, life-support systems in the biosphere might be stressed beyond its carrying capacity (9). This is a troublesome issue in the light of the finding that something like 40 percent of the terrestrial photosynthetic productivity is currently used, wasted, or diverted (10).

Third, exacerbation in international relations is likely if there is a three-fold growth in population in the 40 low-income countries where poverty is endemic, while the individual consumption of goods and services in 25 high-income countries increases 2 1/2 times to a level more than 40 times that in the low-income countries.

These issues are beginning to engage the serious attention of scientists as they explore the three questions: What kind of world do we have? What kind of world do we want? What must we do to get the world we want? (3) In brief, we now have a world that is unsustainable, inequitable, and unstable. It is unsustainable in the sense that exponential demographic and economic growth cannot continue indefinitely without jeopardizing the life-sustaining local, regional and global environments. It is inequitable in that abysmal poverty and excessive consumption exist virtually side by side. It is unstable in that the demographic and economic asymmetries are widening in a manner that is all but certain to heighten the tensions that now exist between the high- and the low-income countries.

We want a world that is sustainable, equitable, and stable. The things that we must do to get the world we want include the imperatives of (i) reducing poverty everywhere, (ii) stabilizing world population, (iii) minimizing the environmental impact of modern technology so that necessary economic development can proceed without further environmental deterioration, and (iv) re-examining patterns of excessive personal consumption.

An historic conference of over a hundred nations and several hundred nongovernmental organizations in Rio de Janeiro in June 1992 provided a point of departure for the world to address these issues and to begin the response to these imperatives (11). Great emphasis was placed there on the concept of *sustainable development*. This has been described as "...the ability to make development sustainable — to ensure that it meets the needs of the present generation without compromising the ability of future generations to meet their own needs." (12)

The Earth Summit and associated nongovernmental Global Forum constituted a "hinge on history" that presents a special challenge to scientists (13). It is useful to try to distill from the 2500 actions that were adopted at the Earth Summit as an agenda for the 21st century, those central tasks that should command early attention of scientists in fulfilling their responsibilities for science and public policy. In my view, they are four:

(i)   To deepen our understanding of the manner in which the great global physical, chemical, biological and social systems interact to regulate planet Earth's favorable environment for human life and ascertain what limits on human activity may exist.

(ii) To stabilize world population, with particular attention to low-income countries.

(iii) To transform the energy- and technology-driven economy that produces and consumes goods and services into one that is environmentally benign.

(iv) To re-examine societal goals, particularly in high-income countries, with the intent of giving relatively more emphasis on the quality of life and on meeting essential needs and legitimate aspirations, than on acquisition of evermore goods and services.

A promising initiative in the task of deepening our understanding of the interaction of natural planetary systems and human activity is found in the International Geosphere-Biosphere Program of the International Council of Scientific Unions (14). A lineal descendent of the International Geophysical Year and the International Biological Program, it is now moving toward embracing the social system.

Stabilization of population requires the classical demographic transition from high to low birthrates and death rates (Figure 5). Many factors are involved in this transition. They include education, economic development, elevation of the role of women in society, and adoption of methods of family planning.

Transformation of the economy requires an environmental transition. Figure 6, adapted from a suggestion by T. Graedel and R. Socolow (15), suggests a relationship between environmental impact and what in this instance has been designated as *sustainable human development*. This concept is the mature formulation of the idea of sustainable development that was elevated into public discourse by the Earth Summit. Sustainable human development embraces the economic growth explicit in sustainable development, but reached beyond economics to include enlargement of personal choices by access to meaningful employment, health, education, and culture. It also includes the inalienable right of each individual to political and religious freedom and personal security, with all of the attendant individual responsibilities.

In the early stages of sustainable human development, primary emphasis is on rapid growth of the energy- and technology-driven economic system. Exploitation of natural resources with little regard for environmental impact characterize this stage. As these impacts increase and become troublesome, measures are adopted to alleviate them. These measures include minimization of energy inputs and material throughputs (16) and major changes in consumption patterns.

The generic path is shown by the curve marked *a*. Some nations (e.g., United States) adopt strong measures to reduce environmental impact. Their path is shown by the curve *b*. Other nations (e.g., central planning countries) are slow to adopt these measures. Their path is shown by the *c* curve. Still others (e.g., low-income countries) are anxious to bypass the peak of environmental impact by drawing upon the experience of nations that have passed that peak. Their paths are shown by the *d* and *e* arrows. These shortcuts require new modes of partnership between high-income countries and low-income countries. The need for this was a recurring theme at the Earth Summit.

Re-examination of societal goals will be the most difficult and sensitive task in preparing for the 21st century. New knowledge of human behavior will be required. Profound changes in our ways of thinking (metanoia)[1] will be necessary.

Major renewal and innovation must be initiated in the institutions through which individuals bring their wisdom and imagination to bear on the human prospects as we cross the threshold into the first century of the next millennium.

---

[1]"A Fundamental transformation of mind and character," p. 1420, Webster's Third International Dictionary, G. & C. Merriam, 1976.

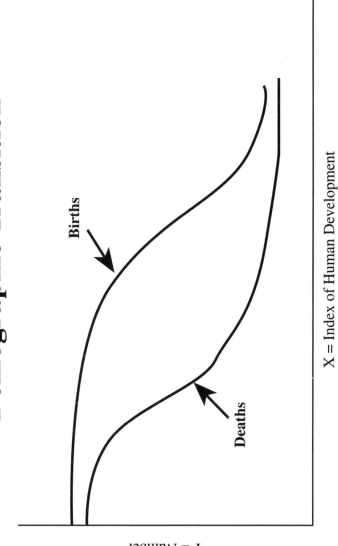

Demographic Transition

Births

Deaths

Y = Number

X = Index of Human Development

Figure 5

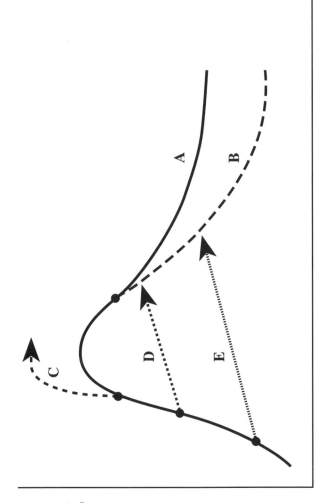

Figure 6

Central to the responsibility of the scientist is the conclusion that the world is on the threshold of a new era characterized by a regime of knowledge in which the generation, integration, dissemination, and application of knowledge will be a prerequisite for human progress.

The institutional requirements were addressed with prescience by the World Commission on Environment and Development in their 1987 report *Our Common Future* (12). They remarked that "A major reorientation is needed in many policies and institutional arrangements at international as well as national levels because the rate of change is outstripping the ability of scientific disciplines and our current capabilities to assess and advise. ...A new international program for cooperation among largely nongovernmental organizations, scientific bodies, and industry groups should therefore be established for this purpose."

There was an immediate response at the Second World Climate Conference in Geneva in 1990 in the call "to create a network of regional interdisciplinary research centres, located primarily in developing countries, and focusing on all the natural science, social science, and engineering disciplines required to support fully integrated studies of global change and its impact and policy responses ... and (to) study the interaction of regional and global policies." (17)

This suggestion was developed further as an integral part of the International Geosphere-Biosphere Program of the International Council of Scientific Unions in the concept of a global SysTem of regional networks for Analysis, Research, and Training (**START**) now beginning to be implemented (18). The Carnegie Commission on Science, Technology, and Government has proposed the creation of a consortium of donor agencies "To strengthen and interrelate the worldwide capabilities for environmental research, especially as they apply to development." ...and for "Facilitation of national collaborative networks." (19). The consortium would be an international Consultative Group for REsearch on ENvironment (**CGREEN**) patterned with modifications after the highly successful Consultative Group on International Agricultural Research (CGIAR) (20).

In 1992, eleven countries of the Americas created the Inter-American Institute for Global Change Research (21). It is an array of research institutions chartered "...to conduct and sponsor research on global change processes of special importance and, in some cases, unique to, the Americas. ...to expand the frontiers of knowledge and serve as an effective interface between science and the policy process."

It is clear that a major international institutional initiative is underway. It will involve an interdisciplinary effort of unprecedented scale, embracing all of the sciences, engineering, and the humanities. To link effectively the knowledge regime with the formulation of policy in the public and private sectors public will require forging new modes of communication and partnership among four key sectors of society:

(i)   governments responsible for the commonweal,

(ii) business and industry, responsible for the production and distribution of goods and services in an environmentally benign manner,

(iii) academia and the research community as the principal custodians of knowledge, and

(iv) nongovernmental organizations that bring together individuals with similar interests, common aspirations and shared values as catalysts for progress (foundations, scientific and professional societies, religious organizations and volunteer environmental and developmental groups).

The institutional "reorientation" that is imminent will bring development of the critically important knowledge base to the local and regional level where decisions are made most effectively. It can foster a dynamic and creative interaction among science, technology and society. This is a prerequisite for successful pursuit of a new world vision. Articulation and promulgation of that vision is absolutely essential for creating the individual and collective *will* to take action during this decisive decade in order to assure an attractive human prospect as we enter the first century of the third millennium.

That vision is a global society in which the essential human needs and an equitable share of human aspirations and wants can be met by everyone in present and future generations, while maintaining in perpetuity a healthy, physically attractive and biologically productive environment for all. This is a vision that has the potential of unifying and galvanizing a world society wracked by fratricidal conflict, plagued by social and economic injustices, and existing with the knowledge that the present arsenal of nuclear weaponry is sufficient to obliterate civilization in a matter of days or hours.

What, then, is the responsibility of the scientist? This is a highly personal matter. The answer is influenced by circumstances peculiar to each individual. If the aggregate of individual reactions by many scientists is to be effective in helping to chart an attractive course for society, a few essential steps are indicated.

First, a kind of metanoia is in order. It can begin only with a commitment by the scientist to become informed on the general nature of the trajectory along which the generation of knowledge is propelling society. This awareness should be broadened by reflective exploration of the role that the other three dimensions of knowledge play in determining that trajectory.

This thought process will lead the individual across the boundaries of the discipline in which competence has been acquired. Disciplinary roots, however, must be treasured and nourished. The effectiveness of interdisciplinary approaches is dependent on the vitality of the disciplines that are involved in that activity. "Tithing" of one's time and talent offers an approach to preserving disciplinary competence while enriching an interdisciplinary endeavor (4).

Second, it is clear that more than fostering an intellectual outreach is required to fulfill the responsibility of the scientist. The era is drawing to a close when the scientific enterprise can be considered apart from the society in which it is embedded (22, 23). Action is also indicated. Individual action can be made directly through any one of several institutions in which the scientists is involved: university, corporation or other business enterprise, scientific and professional society, church, and local, state, and national governmental bodies. Here, too, deliberate tithing of time and talent is a way to participate in the formulation of intertwined science and public policy.

As a closing thought, the hundred thousand scientists in Sigma Xi could be a potent force in responding to the "attractive opportunities and formidable problems" mentioned earlier as a challenge to society — and to the scientist — in the 21st century. This is a very extraordinary ethical issue that deserves the attention of each one of us.

References

(1) Carson, R., *Silent Spring*, Houghton Mifflin, 1962.

(2) Boyer, E., *Scholarship Reconsidered*, The Carnegie Foundation for the Advancement of Teaching, Princeton, 1990.

(3) University Colloquium on Environmental Research and Education, Proceedings, September 24-26, 1992, Sigma Xi, The Scientific Research Society, Research Triangle Park, NC.

(4) *Global Change and the Human Prospect: Issues in Population, Science, Technology, and Equity*, Forum Proceedings, November 16-18, 1991, Sigma Xi, The Scientific Research Society, Research Triangle Park, NC, 1992.

(5) World Bank, *World Development Report 1990: Poverty*, Oxford University Press, 1990.

(6) World Bank, *World Development Report 1992: Development and the Environment*, Oxford University Press, 1992.

(7) Sinyak, Y., and Nagano, K., *Global Energy Strategies to Control Future Carbon Dioxide Emissions*, Status Report, SR-92-04, International Institute for Applied Systems Analysis, Laxemburg, Austria, October, 1992.

(8) *Climate Change, The LPCC Scientific Assessment*, J.T. Houghton *et al* (eds.), Cambridge University Press, 1990.

(9) Daily, G., and Ehrlich, P., "Population, Sustainability, and Earth's Carrying Capacity," *BioScience*, November, 1992.

(10) Vitousek, P., *et al*, "Human Appropriation of the Products of Photosynthesis, *BioScience* 36, 368-373, 1986.

(11) Haas, P.M., Levy, M.A., Parson, E.A., "Appraising the Earth Summit: How Should We Judge UNCED's Success?", *Environment*, 34 (8), pp. 7-11 and 26-33, October, 1992 (see also "A Summary of the Major Documents Signed at the Earth Summit and the Global Forum", the same issue of *Environment*).

(12) World Commission on Environment and Development, *Our Common Future*, (The Brundtland Report), Oxford University Press, 1987.

(13) Malone, T., "The World After Rio," *American Scientist*, December, 1992.

(14) The International Geosphere-Biosphere Programme: A Study of Global Change, ICSU/IGBP Secretariat, Royal Swedish Academy of Sciences, Stockholm, 1990.

(15) Private communication, September, 1992.

(16) Ausubel, J., "New Technologies for a Sustainable World," Hearings before the Subcommittee on Science, Technology and Space of the Committee on Commerce, Science and Transportation, U.S. Senate, Washington, DC, June 26, 1992.

(17) Jaeger J., and Ferguson, H., *Climate Change: Science, Impact and Policy*, Proceedings of the Second World Climate Conference, Cambridge University Press, 1991.

(18) Eddy, J., *et al, Global Change System for Analysis, Research and Training* (START), Report No. 15, International Geosphere-Biosphere Program, International Council of Scientific Unions, Boulder, CO, 1991.

(19) Carnegie Commission on Science, Technology, and Government, *International Environmental Research and Assessment: Proposals for Better Organization and Decision Making*, New York, July, 1992.

(20) Baum, W., *Partners Against Hunger*, World Bank, 1986.

(21) *IAI/IC Newsletter, Issue 1*, (Ereno, C.E., Ed.), Servisio de Hidrografia Naval Av. Montes de Oca 2124 (1271), Buenos Aires, Argentina, 1992.

(22) President's Council of Advisors on Science and Technology, *Renewing the Promise: Research-Intensive Universities and the Nation*, Washington, DC, December, 1992.

(23) Brown, G., (Chair), *Report of the Task Force on the Health of Research*, Committee on Science, Space and Technology, U.S. House of Representatives, Washington, DC, July, 1992.

Paradoxical Strife: Science and Society in 1993

J. Michael Bishop

We live in an age defined by science, when many of nature's great puzzles have been solved. Despite transcendent achievement, however, science now finds itself in paradoxical strife with society. We have reached a time in the life of our nation when it has become necessary to reconsider the ways in which science proceeds, the benefits and stresses it brings, and the means by which it can be sustained.

I will address these issues mainly from the perspective of biomedical research, the area of my expertise. But I believe that my remarks will illustrate problems and opportunities found throughout science. As a point of departure, I will use developments in human genetics that dramatize the ways in which science interpenetrates and even destabilizes society (1).

Genes and Disease

We have reached the molecular level in our pursuit of human disease. Most if not all of our great maladies have their groundings in molecular failure, deriving often from disorder in the structure and function of genes.

First, there are numerous hereditary diseases that arise from defects in single genes. We have found the genetic lesions in some of these diseases, we are closing fast on the lesions in others. The scope of the problem is substantial: each one of us is likely to be the carrier of at least thirty recessive genetic ailments; new reports of single-gene defects in human beings reach the medical literature several times a month; more than one-hundred genetic diseases afflict our retinas alone, most of which are probably caused by defects in single genes.

Admit more complex ailments to the repertoire, however, and the scope becomes all-encompassing. Even our most common afflictions are rooted in our genetic dowries; examples include atherosclerosis, hypertension, cancer, allergies, diabetes, Alzheimer's Disease, schizophrenia, manic-depressive disease and infections. If you doubt this, ask yourself why we are not all equally susceptible to these ailments, whatever their immediate causes.

Take infectious disease as an example that might not come first to mind. Individuals vary greatly in their susceptibility to most infections, a variation whose bedrock is our genome. Only the most recent microbial invaders of our species enjoy uniform success as pathogens when they infect us - the deadly efficiency with which HIV causes AIDS is the latest reminder of this grim principle. Left to their own devices, the forces of evolution eventually reshape the genomes of host and microbe to a less destructive interplay.

Our struggle against AIDS will probably be won by means of prevention, a time-honored strategy in our dealings with pestilence. But the struggle would be easier if we understood the rules that govern our response to the AIDS virus, and those rules are written in our genome. Once we can make a profile of a person's genetic predisposition to disease, medicine will finally become predictive and preventive.

Genetic Testing

What are the prospects for forging a predictive and preventive medicine based on genetic testing? More than 25 of the single-gene defects in human beings can now be diagnosed by molecular techniques, and this is only a beginning. The visionaries among us foresee a time when we will be able to detect even the most multifactorial susceptibilities to disease and to take preventive action.

I once thought that this prospect would prove illusory. But now it seems only a matter of time before we will be able to test for all substantive genetic predispositions to disease, including those that are multigenic. For example, testing for genetic predispositions to cancer is rapidly becoming a reality; hypertension, atherosclerosis and diabetes have become realistic targets for the longer term.

The task will be large. We might well examine more than 20 million individuals each year. But place this in perspective: five billion medical laboratory tests are performed in the United States annually; genetic screening would increase that burden by perhaps 5 to 10 percent; it would not be an impossible or even impractical task should we choose to do it; we presently perform more than 2.5 million tests annually for fetal trisomy alone.

Genetic Testing and the Prevention of Disease

We know from experience that detection of genetic maladies can influence the prevention of disease. Since the introduction of prenatal screening for Thalassemia, the frequency of newborns with this disease has declined by more than 60 percent in Italy, a country dominated by Catholicism and its proscriptions against both contraception and medical abortion. Unhindered by such proscriptions, the Greeks of Athens have virtually eliminated Thalassemia from new births.

Current estimates hold that more than 60 percent of all deaths in the United States are premature, owing to inadequate prevention. The advent of genetic testing would offer the chance to change that figure for the better. The change would require preventive measures that go far beyond the avoidance and termination of pregnancy, intruding on personal behavior to an extent that many would find objectionable, projecting physicians into an arena that they have traditionally neglected.

There will be economic issues to resolve. Will the screening be cost-effective? Will it improve case-ascertainment sufficiently to justify its use and expense? And will we be able to act on the information it provides, or will we be stymied again by our inadequacies in the prevention of disease, inadequacies that stem mainly from failures of personal and political will? If all this comes to pass, it will represent an over-arching change in the practice of medicine, disruptive of our established ways, threatening great strife if implemented before the medical profession and the general public have been properly prepared.

Genetic Testing and Social Strife

The opportunity for prevention through genetic testing creates diverse difficulties — vexing, compounded by a dense ethical haze, unexpected. More than a decade ago, Sweden implemented screening of children for deficiency of an enzyme known as alpha-1-antitrypsin. The objective was to caution affected individuals not to smoke, in order to avoid otherwise certain emphysema. But what seemed to be a harmless and helpful message proved to be so great a stigma that the screening program was discontinued.

The need for confidentiality in genetic testing has inspired great anxiety in the United States. If given access to the genetic profile of an individual, potential employers and insurers would be unlikely to ignore any warning signs written there. Indeed, there is concern that genetic testing might become a requirement for insurance and employment. Individual states have begun to enact legislation that prohibits access to the results of genetic testing without consent of the individual. But consent can be coerced by social and economic means: the problem remains unsolved.

Some observers worry that genetic testing will refuel the call for eugenics. But surely our society can find the wisdom and means by which to deal with that unwelcome and oft-repudiated prospect. It is, in fact, all too easy to underestimate the decency and courage of our fellow citizens.

Late last year, *The San Francisco Chronicle* carried a feature article about genetic testing, focused on cystic fibrosis (2). In particular, the article told of a couple who learned first that both were carriers of the recessive defective gene responsible for the disease, then that the twins they had recently conceived were afflicted with the disease itself. After an agonizing debate, the couple decided to continue the pregnancy.

Here is what the mother had to say during the last trimester of her pregnancy. "We will be sad parents, but parents who have had the chance to grieve for their loss. Parents who are informed and able to provide [their children] with a healthy and loving environment. Am I glad that I had genetic testing? Yes. I have a sense of what I am facing and I am ready to do my best for my children. Knowledge is indeed power...."

It is not eugenics that I fear. Instead, I share with Nobel laureate Max Perutz a different anxiety, the prospect of "a democracy so scared of science that it might accede to the shrill demands and intimidation by those who want termination of pregnancies to be banned, together with genetics and all its works (3)."

In the United States, executive, legislative and judicial action have all been used to obstruct the use of biomedical science. We have just completed a decade of such obstruction. The change of Federal administrations has moved us into a welcome period of legal respite. But the future remains hostage to judicial whim and electoral fortunes.

Frustration and the Funding of Research

While we struggle to balance the promise of science with social conflict, we must confront another challenge: disquiet over the stewardship of science. Symptoms of this disquiet include indignation at abuses of indirect costs and Congressional inquiry into research fraud. But these are parochial symptoms only. The main issue runs broader and deeper.

One of my predecessors in this lectureship, Gerald Holton, anticipated the problem when he described the views of the philosopher, Imre Lakatos (4). Lakatos imagined that it should be possible to distinguish between two generic sorts of research: progressive and degenerating (with elementary particle physics being an example of the latter). He proposed this distinction as a suitable basis for decisions on both the publication and the funding of research.

Enter Congressman George Brown and his Task Force on the Health of Research. Congressman Brown has been a durable and thoughtful friend of science, but now he is toying with alarming departures, the makings of great mischief (5). He complains of what he calls a "knowledge paradox": a parallel rise in fundamental knowledge on the one hand, societal crisis on the other. He implies that the two curves should be reciprocals of one another — that as science progresses, the problems of society should diminish. He wonders why science has not contributed more to the achievement of national goals and suggests that we may have to change the ways in which we identify research for funding.

Those thoughts from Congressman Brown are trouble enough. But his Task Force has truly taken the bit in their teeth. Here are some of their suggestions.

(i) Congress should exert greater control over the choices of research to be funded — in particular, by making The Office of Science and Technology Policy (OSTP) a "command center for implementing and evaluating all major research policy decisions" (6) (the OSTP just happens to fall under the jurisdiction of Congressman Brown's Committee on Science, Space and Technology).

(ii) Research should be addressed more immediately to "current political, economic and societal pressures" (7); research programs "[should be linked] explicitly to goals in a manner that would optimize the policy-relevance of the research" (8). Implicit in this suggestion is the assumption that it is, in fact, possible to determine in advance which research is most necessary for a given national goal. But how many scientists would concede that assumption?

(iii) Legislative mandates should be used to determine how research is evaluated (9). The inherent excellence of research is not considered a sufficient criterion for judgment because "[it] does not guarantee policy relevance or potential application to technological development" (10). Thus, "programs that are failing to meet stated goals should be terminated...." By this parameter, I suppose we should now terminate cancer research because it has so far failed to produce a panacea for the disease.

(iv) It may be preferable to discard peer review as we now know it, in favor of block grants and funding decisions by "smart managers" (11).

(v) The "users" of knowledge should have a greater role in evaluating research performance — examples given include members of the business and legal communities, educators, state and local policy planners, public interest groups, journalists, the military, and policy analysts who conduct "research on research " (12).

(vi) The strategic planning of our research has been inadequate: research has been treated as a "black box into which federal funds are deposited and from which social benefit is somehow derived" (13).

Through these suggestions, the members of the Task Force display an utter ignorance of the history and practice of science. They betray expectations that science cannot meet, a misapprehension of its capabilities. They fail to recognize that the motives of policy cannot mandate success in science: the directions of science are ultimately dictated by feasibility — science is the art of the soluble, of the possible, to recall a phrase from Peter Medawar (14). They ignore the substantial strategic planning that has guided both fundamental and applied research in the United States over the past fifty years, and the plentiful results that have redounded to the benefit of society: effective vaccines against more than a dozen viral and bacterial ailments; particle accelerators and all they have spawned; the rapid isolation of the virus responsible for AIDS; the semiconductor, computer and biotechnology industries. These advances did not come from random walks through the vineyards of research.

Above all else, the Task Force misplaces blame: science has long since produced the vaccines required to control most childhood infections in the United States, but our society has not found the political will to properly deploy those vaccines; science has long since sounded the alarm about acid rain and its principal origins in automobile emissions, but our society has not found the political will to bridle the internal combustion engine.

Congressman Brown argues that "we must test the hypotheses that link economic and social benefits directly to research" (15). I had thought those hypotheses were by now well proven.

The members of the task force were congressional staff, few with any substantive credentials in research. They have written a script for mischief. Worse yet, the script is beginning to play out.

(i) There are mounting demands that the National Science Foundation (NSF) redirect its efforts towards technological development, thus tampering with a budget that is on the one hand a bulwark of fundamental research in the United States, on the other hand only 3 percent of Federal spending on research and development — one of the last places from which monies for technological development should be drained.

(ii) Earmarking of biomedical research funds by Congress has become a growth industry, mounting to more than 400 million dollars in biomedical research alone this year (1.2 billion in total research dollars over the past two years).

(iii) Frustrated by the refractory nature of AIDS, activists and political leaders alike have called for a new administrative structure to direct research on the disease, resorting to potentially misleading analogies with the Manhattan Project and the Russian empire ("give us a czar," they cry). The result: a cacophony of proposals and rancorous debate, all in hopes of a quick fix that is not likely to come by any means. The impetus to reorganize AIDS research now seems irresistible. What alarms me most is the vehicle by which the reorganization seems likely to come: a riotous legislative process rather than strategic planning by the well-informed. Someone should take charge before too much damage is done.

What all this turmoil signifies is that science has become a victim of its own success. So much has been accomplished that far more is expected than we can hope to deliver. Why has malaria not been eradicated by now, why is there still no cure for cancer and AIDS, why is there not a more effective vaccine for influenza, when will there be a final remedy for the common cold, when will we be able to produce energy without waste, when will alchemy at last convert quartz to gold? The litany of frustration seems infinite.

The Endangered Future of Research

The complaints from Congress, the assault on the NSF, the increasing manipulation of research funding by legislation — these are alarming developments because they threaten to undermine the health of our research enterprise by misdirecting precious resources. The last decade has witnessed an accelerating erosion of the infrastructure for fundamental research in the United States. If that erosion is not reversed soon, the pace of discovery will necessarily decline, with wide-spread consequences for industry, health care and education. The United States now runs the risk of enervating an entire generation of scientists and of losing its international leadership in the enterprise of discovery.

Leon Lederman, physicist and Nobel Laureate, caught the spirit of the moment well with these recent words: "...something very dark and dramatic is taking place in our universities, a deep sense of discouragement, despair, frustration, resignation, a quenching of the traditional optimism of research scientists" (16). Make no mistake: this is not hyperbole: I have seen and felt the mood of which Lederman spoke. If you doubt my jeremiad, consider three anecdotes.

(i) A recent survey of American neurochemists revealed that almost half of the full professors said they probably would not have entered biomedical research had the funding situation then been as it is now (17).

(ii) Graduate students in genetics at the University of Washington recently polled themselves about their plans for the future. Two thirds of the students indicated that they planned to avoid careers that might be dependent upon grant support from the federal government.

(iii) Two years ago, I addressed science students at the premier high school in San Francisco. The encounter began as an exhilarating experience: eager young faces and intellects, perceptive questions, enthusiasm for my tale of discovery. But it ended on a disheartening note. When I asked the students about their plans for careers in science, I learned once again that you cannot fool all of the people all of the time — especially when they are young, smart and pragmatic.

"Why should we consider careers in basic research," the students asked me, "when we know that scientists are no longer able to get the grant money they need to work." These youngsters were only 16 and 17 years of age, but they too had heard the message.

The mood exemplified by these anecdotes threatens us with a hemorrhage of talent that will undermine not only our ability to perform fundamental research, but our ability to educate future generations. We need to entice more young minds into science, not discourage them with inadequate funding: the scientists we recruit today are both the discoverers and the teachers of tomorrow.

Public Values and the Funding of Research

I believe that the root of the problem is a perversion of public values. The United States now spends more than 600 billion dollars per year on health care, but less than 2 percent of that on research and development in the bio-medical sciences. By contrast, the defense industry spends 15 percent of its cash flow on research and development; pharmaceutical companies, 9 percent (some place the number as high as 15 percent); the aerospace industry, 6 percent; automotive companies, 5 percent; the tire and rubber industry, 3 percent.

In 1992, the United States spent approximately 1.9 billion dollars on cancer research of all sorts, 1.3 billion on AIDS, 730 million on heart disease, a mere 280 million on diabetes. In contrast, our nation spent in excess of 400 billion dollars on military defense, 170 billion dollars on Fords and Chevies (the sum for Japanese imports is too humiliating to be mentioned), $140 billion on "recreational drugs" (all of them presently illegal), 6.9 billion dollars on subsidies so farmers would not grow crops, 4.2 billion on Star Wars research, 1.8 billion on the Nintendo computer game. Surely there is room here to spare for fundamental research of all sorts, for the Human Genome Project and the Superconducting Super Collider - even if the last must be located in Texas.

I took pains to mention the Human Genome Project and the Super Collider because they exemplify a special problem in the funding of science: the threat that we may rob Peter to pay Paul, small science to pay large. With costs running beyond 3 and 8 billion dollars, respectively, the two projects represent the sort of gigantism that is anathema to many biological scientists.

Gigantism has its uses — witness this culinary aphorism from 18th century China: "The cook who works only on a small scale lacks grace. Good cooking does not depend on the size of the dish or the cost of the ingredients." The aphorism echoes what Freeman Dyson has said about big science in this century (18): "We cannot calculate from general principles the optimal size of a scientific project, any more than we can calculate the optimal size of a whale."

Dyson happens to like the Human Genome Project because of its strategic pragmatism. He is less enthusiastic about the Super Collider, urging that we build "several clever accelerators instead of one dumb accelerator" (19). It goes without saying that the clever accelerators will require new ideas before they can be built, and they will be less expensive than the dumb one.

The *New York Times* has offered an even more Solomonic solution: require physicists to moth-ball enough of the existing obsolescent high energy machines to cover the costs of the new one. I am merely reporting this suggestion, not advocating it.

Can we even afford the Super Collider? I am one biologist who says "yes." After all, the device would cost less than a single Sea Wolf submarine, which we have not hesitated to build in multiples. In the words of Steven Weinberg, "without [the Collider], we may not be able to continue with the great intellectual adventure of discovering the final laws of nature" (20). I quote Weinberg because he is an authoritative and eloquent advocate of the Super Collider; but of course, he is also an adopted Texan.

As for the Space Station, that champion Gargantua of the moment, suffice it to say that its proponents now use the prospect of biomedical experiments in space as its principal justification, and that just will not wash: no biologist of my acquaintance believes there is anything we could do on the Space Station that would come close to justifying its 30 billion dollar price tag.

We must move beyond categorical debates over big and small science, to distinctions made on objective criteria, such as intellectual quality and promise, potential utility, and intelligence of design. Steven Weinberg again (21): "Arguing about big science versus small science is a good way to avoid thinking about the value of individual projects." Big science is not inherently evil. But it must be judged with meticulous care and honesty.

Calculating Needs and Returns

How much should we spend on science in the aggregate, how much is needed? I think it is fair to say that no one really knows. No one has yet devised a calculus for funding that science and government can both trust. For the moment, science solicits according to perceived opportunity, government appropriates according to what the traffic will bear. Surely we could improve on that crude formula.

Part of the difficulty is that we have never adequately calculated the return on our investment in fundamental research. Again, it is not clear that anyone as yet knows how to make the calculation. But the available approximations suggest a staggering return. To cite one example from my own purview: it has been estimated that the vaccine against poliovirus now saves the United States billions of dollars every year in costs of health care and losses of productivity, whereas the cost of developing and distributing the vaccine can be reckoned in the mere millions. We need to improve on calculations of this sort and deploy them in design of the Federal Budget. Until we do, science will continue to solicit at an unnecessary disadvantage.

Furthermore, a strictly economic argument fails to encompass the full return. Let Congressman Brown make the point (22): "Basic research represents a uniquely human quest to achieve intellectual and spiritual insight and growth through scientific inquiry." "Particle accelerators, spacecraft, cathedrals and libraries all are essentially similar. They are settings for cultural experience."

I challenge the prevailing view that we have reached the limits of our capacity to fund fundamental research, that a steady state (at best) must now prevail. That view is founded on outmoded priorities and requires close examination as we redirect Federal funds away from weaponry and war.

For the moment, at least, the news from Washington has taken a turn for the better. As part of their investment package, the Clinton-Gore administration is proposing substantial increases in the budgets for the National Institutes of Health (NIH) and the NSF. We of science must now seize the day and make the case that these increases are justified. We should not be reticent: we speak in self-interest, of course, as do all beneficiaries of the Federal patron; but we also speak for one of the noblest endeavors of human kind.

It is not easy to raise claims for science in the face of the social strife that repeatedly assails our nation, strife that makes the limits on science seem a parochial issue. But our nation remains prosperous and generously endowed. If in a fiscal instant, we can find 500 billion dollars to rescue a dubious savings and loan industry, if in a fiscal instant we can find the uncounted billions required to blunder into war with a nation one-fifteenth our size, surely we can find the resources required to secure an enlightened future, for ourselves and for the generations that follow.

Public Altruism and Science

The public may be indifferent to our fiscal plight, but they are not indifferent to our presence among them. For more than five years, my school — the University of California, San Francisco (UCSF) — has been waging a costly and enervating battle for the right to perform biomedical research in a residential area. The opponents: our neighbors, who argue that we are noisome beyond tolerance; that we exude toxic wastes, infectious pathogens and radioactivity; that we endanger the life and limb of all who come within reach — our own lives and limbs included, I suppose, a nuance that seems lost on the opposition.

Two images from the fray will suffice to dramatize our plight: an agitated citizen, suggesting in public forum that the manipulation of recombinant DNA at UCSF engendered the AIDS virus; and an elderly denizen of the neighborhood, declaring over television her outrage that "those people are bringing DNA into my neighborhood."

These views and others of similar ilk fueled a reaction that has stopped the university dead in its tracks and that promises no end of trouble for the foreseeable future. A city official recently called the episode "one of the most tragic" in the history of San Francisco — an overstatement, perhaps, given our history of catastrophic earthquakes.

The heart of the problem is a renunciation of public altruism. The Supreme Court of California spoke to the issue of altruism, in its ruling on the strife in San Francisco. While chiding the university for not having done its homework well enough (the initial Environmental Impact Report was deficient), the court said that it did not wish to "shackle the scientific imagination" with unrealistic standards; the court acknowledged the inherent unpredictability of research and its hazards; the court even argued that the salutary nature of our mission mitigates the unpredictability, indeed, mitigates the hazards themselves — which the court recognized as miniscule beyond calculation. In the words of Bernard Davis (23): "...mankind has always faced risks, whether in exploring uncharted territories or trying unfamiliar foods. If our recent success in conquering many malign forces of nature now leads us to seek the security of a world free from novel hazards, and if we forbid exploration of the new kind of unknown territory opened to us by science, we shall not only be condemning ourselves to remain subject to all the present, still unconquered risks; we shall be crushing one of the most admirable expressions of the human spirit."

The Deepest Malady

Resistance to science is born of fear. Fear in turn breeds on ignorance. And ignorance is our deepest malady. The problem is before us daily in the United States: in the evidence of woeful scientific literacy among our populace; in the failures of our elementary and secondary schools to teach science well (or at all); in the rancorous disputes over the place of science in the general curricula of our undergraduate colleges; in the bewilderment of laborers, accountants, lawyers, poets, politicians, even physicians, when they look on the body of science. The consequences are dire.

(i) According to recent polls, more than 90 percent of U.S. citizens still believe that, even if evolution accounts for the origin of species, it must have been guided by a supernatural hand, not by the play of natural selection.

(ii) The previous Secretary of the Interior of the United States denied the evolutionary origin of human life (24): "God created Adam and Eve, and from there, all of us came. God created us pretty much as we look today."

(iii) In California, religious zealots maintain a steady drumbeat of pressure to restore the mandatory teaching of creationism to our public schools: the world may be round, gravity may be real, the earth may circle the sun — these oddities are at last commonly accepted (although not necessarily commonly known); but evolution never happened.

(iv) Campaigns against the use of experimental animals in medical research have permeated our public schools, sometimes conducted by the teachers themselves, and Congressional legislation on the use of animals threatens to become increasingly restrictive — several years ago, a bill from my then Congressional representative described medical experiments with animals as "cruel and inhumane."

(v) For eight years (at least), presidential decisions in the U.S. were apparently influenced by the nonsense of astrology.

Do we of science even understand one another? I learned recently of a Russian satellite that gathers solar light to illuminate large geographical areas in Siberia. "They are taking away the night," I thought. "They are taking away the last moments of mystery. Is nothing sacred?" But I went on to wonder what physicists must think when hearing that biologists hope to decipher the entirety of the human genome and then, perhaps, to recraft it, ostensibly for the better.

Some years ago I wrote an article about cancer genes for *Scientific American*. I labored mightily to make the text universally accessible: I consulted students, journalists, laity of every stripe. When these consultants had all approved, I sent the manuscript to my brother, a solid state physicist of considerable merit. One week later, the manuscript came back, with a message: "I have read your paper and shown it around the staff here. No one understands much of it. What exactly is a gene?"

Robert Hazen and James Trefil have numberless such anecdotes, which they use to dramatize their advocacy of general science education (25): 23 geophysicists who could not distinguish between DNA and RNA; a Nobel Prize-winning chemist who had never heard of plate tectonics; biologists who thought that string theory has something to do with pasta.

We are amused by these circumstances: we should also be troubled. If science is no longer a common culture, what can we rightfully expect of the laity by way of understanding?

Consider *Lorenzo's Oil*. Lorenzo is an unfortunate child who suffers from a rare hereditary disease known as adrenoleukodystrophy (ALD). The disease destroys the myelin sheath of nerve fibers, cripples numerous neurological functions, and in the form affecting Lorenzo, leads eventually to death.

Offered no hope by conventional medical science, Lorenzo's desperate parents scoured the medical literature and turned up a possible remedy: administration of two natural oils, monounsaturated fatty acids known as erucic and oleic acid. In the face of skepticism from physicians and research specialists, Lorenzo has been given the oils and, in the estimation of his parents, has ceased to decline, perhaps even improved marginally.

The story of Lorenzo is told by a film entitled *Lorenzo's Oil*, which has earned one of its stars, Ms. Susan Sarandon, nomination for an Academy Award. The film portrays the treatment of Lorenzo as a success, with the heroic parents triumphant over the obstructionism of medical scientists. What the film leaves unspoken is that a large number of children with ALD have received the oils in controlled studies, without showing any convincing improvement in their clinical course. The course of Lorenzo's disease to date is little different from that of many other children with the same affliction. Parents of other children suffering from ALD have tried the oils: some claim success, others are bitterly disappointed (26).

The film is deeply troubling in its portrayal of medical scientists as insensitive, close-minded and self-serving; and in its impatience with controlled studies as needlessly wasteful of time — an echo of the outcry from AIDS activists over the past decade. Paradoxically, the film seems to endorse the legitimacy of science: Lorenzo's parents turn to the obscure research literature and to biochemical reasoning to find their remedy. The villain of the story is not science itself but scientists themselves, seen through the eyes of two despairing and intelligent human beings. One line spoken by Lorenzo's father late in the film encapsulates the argument: "These scientists have their own agenda and it is different from ours." Here is a warning we cannot take lightly (27).

As if on cue, isolation of the gene ostensibly damaged in ALD has just been reported (27). Thus, the exact biochemical defect responsible for the disease is known at last (and could not have been predicted from what was known before). The stage is set for the development of decisive clinical testing and therapy, although the latter may still be long in coming.

I recently found myself in conversation with a computer engineer who had seen *Lorenzo's Oil*. To my dismay, this individual accepted the viewpoint of the film without question, argued that the controlled clinical studies of the oils have no greater validity than the one anecdotal experience with Lorenzo, and assailed the biomedical research community for dragging its feet in the exploration of human disease. To this individual, elucidating the pathogenesis of human disease is no different from designing new software and ought to be conducted along the same lines.

Would that biology were so tractable! I left the conversation once again reminded that the various clans of science do not understand one another, neither in motives, nor means nor substance. And I was caused to wonder whether the instinctive skepticism so vital to the ethos of science has been omitted from the education of computer scientists.

Myth and Funding

Lorenzo's Oil conveys three "myths," identified by the bioethicist Arthur Caplan (28). First, that "cures can be found if only bureaucracy and red tape will get out of the way"; second, that "perseverance, hard work and love can conquer any ailment"; and third, that "mainstream science is indifferent" to the suffering of patients and their families, choosing instead any course that will sustain hegemony and privilege.

The same myths (for they are indeed myths, not reality) have helped fuel the strident complaints of AIDS activists against the biomedical research enterprise. Larry Kramer provided us with a recent example. Mr. Kramer is a playwright and formidable activist in the struggle against AIDS. Writing in the *New York Times* last year, he complained bitterly that science has yet to produce a remedy for AIDS (29). Kramer placed much of the blame on the NIH, which he characterized as a "research system that by law demands compromise, rewards mediocrity and actually punishes initiative and originality."

I cannot imagine what law Mr. Kramer had in mind, and I cannot agree with his description of what NIH expects from its sponsored research. I have assisted NIH with peer review for more than twenty years. The standards used have always been the same, seeking work of the highest originality, but demanding rigor as well (a demand that some may find vexing, but which I as a scientist cannot compromise — there is too much at stake). I have never knowingly punished initiative or originality, and I have never seen the agencies of NIH do so. Proposals for creative research have always been received with joy. I realize with sorrow that Mr. Kramer is unlikely to believe me.

There are critics like Kramer (some from within the house of science, I regret to say) who seek to replace peer review of research with a less formal and more agile system of their own — recall the "smart managers" of the Brown Task Force. They are wrong. First, because such systems are too easily corruptible. And second, because the one we have now works well, whatever its blemishes. Revision may be in order, but certainly not rejection.

Biomedical research in the United States now represents one of the most successful ventures our society has ever mounted, driving the discovery of usable knowledge at a remarkable rate, bringing us international leadership in the battle against disease and the search for understanding, and earning us the admiration and envy of other nations throughout the world. It is most unlikely that we could have achieved all this if we did business the way Mr. Kramer and critics like him claim.

Larry Kramer's disenchantment with the organs of science carries a stern warning. We must close the gulf that now cleaves between science and society, or see our enterprise diminished and great opportunity lost.

Alan Bloom and Science

If science bewilders and disappoints some in their ignorance, it repels others in theirs. A few years ago, Alan Bloom's book, *The Closing of the American Mind*, appeared on coffee tables and best seller lists around the land. The book found a sympathetic readership among many academics, although to my eye, it was primarily a tedious effort to blame rock music on Nietzsche and Kant. I agree that someone needs to take the blame for rock music, but Nietzsche and Kant will not do.

In his book, Bloom likened science to "the absurdity of a grown man who spends his time thinking about gnats' anuses...." "We have been too persuaded of the utility of science," Bloom ranted, "[to perceive] how shocking and petty the scientist's interests appear.... If science is just for curiosity's sake, which is what theoretical men believe, it is nonsense, and immoral nonsense, from the viewpoint of practical men" (30).

Make no mistake: these are not the ravings of a deranged fanatic. Bloom was a distinguished professor at the University of Chicago, and his book carried an admiring introduction from Saul Bellow. Then again, professorial rank is no assurance against derangement, as most any student can tell you.

In recent reading, I learned that Bloom probably owed more than I had realized to Nietzsche, who described university teaching and research as "...[a] molish business, the full cheek pouches and blind eyes, the delight at having caught a worm, an indifference towards the true and urgent problems of life"

(31). These sentiments resonate through the report of the Brown Task Force, the laments of Larry Kramer, the tragedy of *Lorenzo's Oil*, the views of Professor Bloom.

I am reminded of Pedro Guerrero. The former Los Angeles Dodger and St. Louis Cardinal once complained that he is misunderstood by the public because "newspapers write what I say, not what I mean" (32). Could it be that Professor Bloom did not mean what he wrote? Would he have wittingly demeaned the great quests of natural science, such as the search for a Grand Unifying Theory of matter, the exploration of our origins in evolution, the dissection of how the brain engenders mind, the explication of how a single cell becomes the glory of the human organism? Those who find no philosophy here, no poetry, no human perspective — they are ignorant or insensate.

Scientists and Public Interest

Fear, bewilderment, disdain — these are all opponents science must best. And there is one other, which is now current: mistrust. For several agonizing years, a subcommittee of the United States Congress has been investigating whether Professor David Baltimore of Rockefeller University was party to fraud in work that he and others published in the journal *Cell.*

In truth, no participant in the investigation has ever accused Baltimore himself of fraud. Indeed, it is difficult for those of us who watch from the outside to know if fraud was done, since none of the evidence has been formally released, and crucial forensic evidence may never be made public. The United States Attorney in Baltimore, Maryland, had a look at the evidence and said "no case." And the principal conclusions of the original paper still stand, unchallenged by any experimental fact.

The resources deployed in the investigation have been both intimidating and ludicrous: an aggressive Congressional staff with seemingly limitless powers and resources for investigation; a whistle-blower borrowed for consultation from the staff of the NIH, even though he had no semblance of expertise in the research at issue; even agents of the United States Secret Service, who spent many months and many more taxpayers' dollars examining subpoenaed laboratory note books for evidence of falsification.

Suspend for the moment whatever opinion you might have formed about the incident. Is Congress the venue, is Congressional investigation the manner in which the veracity of research and the misconduct of scientists should be explored? If our Federal patrons are discontent, can we of science not answer with a better means? At stake is the very ethos of science: the robust counterpoise of success and failure, of error and correction, of mutual trust and lively criticism, by which we proceed. Our critics have failed to apprehend or appreciate this ethos, we have failed to teach and to defend it.

If science were shot through with corruption, as some of our critics in Congress seem to believe, how could we have achieved or maintained the dizzying pace of discovery that has characterized the recent decades in research? Each of us is utterly dependent upon the truthfulness of our colleagues in science: each of us builds our discoveries on the work of others: if that work is false, our constructions fall like a house of cards and we must start all over again. The success of science has always been built on integrity, and that success has never been greater than in our age. The public is ignorant of the formula by which science advances and, hence, is easy prey for our critics.

Science and Public Education

In his Jefferson Lecture of 1972, Lionel Trilling complained that no "successful method of instruction" had been found that could give a comprehension of science to "those students who are not professionally committed to its mastery and especially endowed to achieve it" (33). The problem perceived by Trilling remains with us today: perplexing to our educators, ignored by all but the most public-minded of scientists, bewildering and vaguely disquieting to the general public.

In the face of this great problem, our nation has allowed the means of primary and secondary education to deteriorate. Our teachers are neglected, disrespected, inadequately compensated and improperly prepared. Many of our children attempt to study in the midst of physical squalor and personal decay. We can expect little improvement in how our youth learn until we have changed all of that. The change will require great resolve: we have allowed the deterioration to run very deep.

When I visited that high school in San Francisco two years ago, I was met outside the front door by a delegation of students — a gesture that struck me as unnecessary for the arrival of a mere adult. I soon understood their purpose: they had come to apologize in advance for the deplorable state of the halls within, hoping to blunt the adverse impression they were sure I would gain once I entered.

In that moment on the front steps, I felt indicted of grave neglect — as a parent, as a citizen and taxpayer, as an educator. I cannot repeal the indictment: none of us can. We simply must do better. If we do not, the materialism that erodes at our culture will eventually undermine all learning. And the generations to come will damn us for it.

We of science can no longer leave this problem for others to solve. Indeed, it has always been ours to solve, and all of society is paying for our neglect in precious coin. In the words of Gerald Holton in his McGovern Lecture: "Science is, and must be, part of the total world view of our time (34)." " ...persons living in this modern world who do not know the basic facts that determine their very existence, functioning, and surroundings are living in a dream world. Such persons are, in a very real sense, not sane (35)." "We...

should do what we can, or we shall be pushed out of the common culture. The lab remains our workplace, but it must not become our hiding place" (36).

CODA

The enterprise of science embodies a great adventure: the quest for understanding in a universe that may be "infinite in all directions, not only above us in the large but also below us in the small" (37); the quest for understanding on behalf of life, whose great gift to our planet is diversity, but which remains a scarcely kindled glow in the immensities of our universe.

We have begun the quest well, by building a science of increasing power that can illuminate all that is living. In consequence, we are admired, but also feared, mistrusted, even despised; we offer hope for the future, but also moral conflict and ambiguous choice; the price seems large, but pales in comparison to what it would cost to deny the future. From the American essayist, Annie Dillard: "[Who can read] what the wind-blown sand writes on the desert rock? I read there that all things live by a generous power and dance to a mighty tune; or I read...that all things are scattered and hurled" (38). Will we live by a generous power and dance to a mighty tune, or will we be scattered and hurled?

References

1. The description that follows is based mainly on data and arguments found in N. A. Holtzman, 1989, *Proceed with Caution - Predicting Genetic Risks in the Recombinant DNA Era*. The Johns Hopkins University Press, Baltimore.

2. "Mother-to-be's painful choice," in the *San Francisco Examiner*, November 8, 1992.

3. M. Perutz, in private conversation.

4. G. Holton, 1986. Niels Bohr and the Integrity of Science. *American Scientist* 74: 237-243.

5. G. E. Brown, Jr., 1992. Rational Science, Irrational Reality: A Congressional Perspective on Basic Research and Society. *Science* 258: 200-201.

6. Report of the Task Force on the Health of Research, 1992: Chairman's Report to the Committee on Science Space and Technology, U. S. House of Representatives, 102nd Congress, 2nd Session, p. 14. U. S. Government Printing Office, Washington, D. C.

7. Ibid., p. 5.

8. Ibid., p. 11.

9. Ibid., p. 4.

10. Ibid., p. 12.

11. Ibid., p. 17.

12. Ibid., p. 13.

13. Ibid., p. 11.

14. P. Medawar, 1967. *The Art of the Soluble*. London, U.K.: Methuen.

15. George E. Brown, Jr., op cit, p. 201.

16. For documentation of this mood, see L. M. Lederman, 1991, Science: The End of the Frontier? AAAS Supplement to *Science*, pp. 5-19.

17. R. Haglund, 1991. A Familiar Reply: Neurochemists Say It's No Fun Anymore. *J. NIH Research* 3: 24-27.

18. F. Dyson, 1992. *From Eros to Gaia*, p. 9. Pantheon Books, New York.

19. Ibid., p. 26.

20. S. Weinberg, 1992. *Dreams of a Final Theory*, p. 274. Pantheon Books, U.K.

21. Ibid., p. 273.

22. George E. Brown, Jr., op cit, p. 200.

23. B. D. Davis, 1986. *Storm Over Biology - Essays on Science, Sentiment, and Public Policy*, p. 243. Prometheus Books, Buffalo, New York.

24. Former Secretary of the Interior Manuel Lujan, Jr., as quoted by A. Lewis on the Op Ed page of the *New York Times*, May 28, 1992.

25. For example, see R. Pool, 1991. Science Literacy: the Enemy is Us, *Science* 251: 266-267.

26. For a general account of ALD and *Lorenzo's Oil*, see G. Kolata, "*Lorenzo's Oil*: A Movie Outruns Science," *New York Times*, pp. B5 and B8, February 9, 1993.

27. For a more sanguine view of the film, see "Lorenzo Goes to Hollywood" (editorial), 1993, *Nature Genetics* 3: 95-96.

28. G. Kolata, op cit, p. B8.

29. L. Kramer, Op-Ed page, *New York Times*, November 15, 1992.

30. A. Bloom, 1987. *Closing of the American Mind*, p.270. Simon and Schuster, New York.

31. As quoted in H. Williams, University Challenge. *The Times Literary Supplement*, p. 13, January 22, 1993.

32. As quoted in the *San Francisco Chronicle*, May 13, 1989.

33. L. Trilling, 1972. *Mind in the Modern World*, p.14. The Viking Press, New York.

34. G. Holton, op cit, p. 241.

35. G. Holton, op cit, p. 241.

36. G. Holton, op cit, p. 242.

37. Emil Wiechert, as quoted in F. Dyson, 1988. *Infinite in all Directions*, p. 36. Harper and Row, New York.

38. A. Dillard, 1974. *Pilgrim at Tinker Creek*, p. 68. Harper and Row, New York.

The Value of Science at the "End of the Modern Era"[1]

Gerald Holton

The invitation, and the challenge, to speak to you on the august and embattled subject of "Science and Values" filled me initially with apprehension. As Arthur Rubenstein is reported to have remarked about the music of Mozart, the topic may be too easy for beginners and too difficult for the experts. I would have had the easy task if I were expected to show here, as has been attempted in many volumes before, that science must be valued by society because science claims to be the very embodiment of the classical values, starting with the three primary virtues of truth, goodness and beauty. Making this plausible by demonstrating individual examples would not have been difficult. Thus, science has been widely praised as a central truth-seeking and enlightening process in modern culture — we might call it the Newtonian search for Omniscience. Science also has been thought to embody the ethos of practical goodness in this imperfect world, both through its largely self-correcting practice of honor in science, and through its tendency to lead to applications that may improve the human condition — we might call it the Baconian search for benign Omnipotence. Finally, the discovery of beauty in the structure, coherence, simplicity and rationality of the world has long been held up as the ultimate, thrilling reward for the innovator as well as for the student — we might call it the Keplerian enchantment. That last is of course part of the intense emotional energy behind every individual scientist's work, the counterpart to the so-called cold rationality that seems to frighten the layman.

At any rate, I do not have the easy task today. Most of the optimistic description I have just given was widely taken for granted during recent decades, embodied for example in the famous Vannevar Bush report of nearly 50 years ago, the main driving force of science policy thereafter. And much of it can still be demonstrated convincingly. Despite shortages and other problems that scientists are all too aware of, most of them rarely doubt that the central

---

[1]Portions of this essay are condensations of sections of the forthcoming book, Gerald Holton, *Science and Anti-Science* (Cambridge, MA: Harvard University Press, Fall 1993).

hold of science on the modern mind is secure. However, lately the discussion about science and values has been shifting in remarkable ways, not yet so much at the grassroots, but at the level of the tree tops. Indeed, this symposium is a symptom of that shift. Therefore my task will be the difficult one to try to describe and understand those changes, to put them in historical perspective, and also to gain some inkling of the landscape into which we may be heading, as scientists or as laymen or as general participants of contemporary culture. The aim, then, is clarifying description. There will be too little time here for considering possible remedies; but such remedies tend to become more obvious once one grasps the diagnosis.

We must begin with the notion that at any given time and place, even in a civilization that appears to be in a stable phase, there is an undercurrent of many conflicting ideologies and outlooks. Each of these fervently desires to rise to a position where it would count as the central energizing idea characterizing that particular age and region, and at the same time each is also trying to delegitimate the claims of its main rivals. Especially when the stable phase breaks down, the pandemonium of contrasting voices gets louder, a set of partial victors rises above the rest and then is seen — sometimes more clearly in retrospect — as the ideational embodiment of the new worldview or "sentiment" of that age and place. In that ongoing struggle, from ancient Greece to this day, the scientific conception of the world and how to study it has always played a part, for better or worse, sometimes being the cherished core of the rising or victorious overall worldview, sometimes finding itself embedded in the sinking or defeated one, and then even accused of nourishing, directly or indirectly, a great variety of sins against the better interests of humanity.

Historians of ideas, or of science and technology, have mapped the changing forms of these contrary trends. Wise political leaders, too, have at times watched with apprehension as the net balance of prevailing sentiments has taken a turn, for as Thomas Jefferson said, "it is the manner and spirit of a people which preserve a republic in vigor. A degeneracy in these is a canker which soon eats to the heart of its laws and constitution." Weighty monographs have chronicled how one of the world conceptions, and the scientific position within them, gained predominance over the others for some decades in Western culture, and then gave ground as the overall balance of benignity or distress moved the other way for some more decades, perhaps only to shift back again later still. As to the practicing scientists themselves, at least until fairly recently, they have typically been too busy to pay much attention to this seesaw of history, except to weigh in now and then as promoters of the positive swings, or occasionally to become the victims during the negative ones.

But at our fin-de-siècle, this oscillating spectacle, so engrossing to the scholar, has ceased to be merely the site for their research or amusement. The general balance that had been achieved during the past few decades is changing precipitously before their eyes, and with it a whole range of relations between science and society, hence between science and values. Studying this drama in real time is as fascinating and fruitful for the historian, whose perspective I shall be taking here, as the unexpected explosion of a supernova may be for an astronomer. For what has entered into the equation commanding the up and down motion of the lever of sentiments is an agent, a weight unlike any in the whole history of the rise and fall of the perceived value of science itself. This new agent, and the alchemical forces that have forged it, command closest attention — not merely because of the practical effects they might have on the scientific community, but chiefly for the opportunity that the study of this novel situation may give us a better intellectual grasp on the likely future of science, and also of our culture as such.

I

What is the new agent that has entered? We are all familiar with it, but this very familiarity prevents most of us from seeing the strange power it is having. As in a chemical compound, it consists of three interconnected elements, each of which I shall try to describe. The first element is an assertion, mounting louder and louder over the past few years in books and hundreds of articles, an assertion that has spawned remarkable public hearings, the formation of specific government agencies, university bureaucracies, and even quite a few careers. I refer of course to the widespread assertion that the pursuit of science, to a previously completely unrealized degree, requires us not merely to reassess constantly the safeguards on its ethical practices and uses — of that, there is a long tradition in the scientific community, and that much motivates even our conference here — but that the pursuit of science is, and has been all along, since at least the days of Hipparchus and Ptolemy, thoroughly corrupt and crooked; and that consequently severe measures must be applied to the practice of science from outside.

My favorite example of this assertion is the book by two very influential *New York Times* science reporters, William Broad and Nicholas Wade, which states its intention in the title on the jacket, *Betrayers of the Truth: Fraud and Deceit in the Halls of Science*, and which opens with the unqualified canon-shot of a sentence: "This is a book about how science really works." Going far beyond the need to identify the few rotten apples in any barrel, which the scientific community itself has in fact been the first to recognize, if only for the sake of its own health, this kind of rhetoric has now become commonplace. As this book and its many followers proclaim, the examples of real or alleged misbehavior make out of a few sad cases a litmus test for the whole enterprise; objectivity is a failure, and fraud is shown to be part of the very structure of science. No wonder that the report to Congress by the Congressional Research Service, entitled "Scientific Misconduct in Academia," proposes that, more and more, "the absence of empirical evidence which clearly indicates that misconduct in science is not a problem...suggests that significant misconduct remains a possibility." Among all the imaginable targets to preoccupy those who are charged with timely attention to misconduct damaging our republic, this formulation singles out the conduct of science as being guilty until proved innocent.

Similarly, the Office of Scientific Integrity Review (OSIR) of the Department of Health and Human Services made part of its proposed definition of "misconduct" in science, apart from fabrication, falsification, and plagiarism, "practices that seriously deviate from those that are commonly accepted within the scientific community." The intention here may have been to parallel the way the Supreme Court defined obscenity by reference to the current standards of the local community. However, when it comes to making progress in science, some practices contrary to those common at the time have again and again been the very hallmark of needed innovations — from putting mathematics into physics in the 17th century, to the introduction of quanta, which pained even Max Planck himself, and to the more recent innovation of modern teamwork. The proposed definition of misconduct, with its real potential for mischief, was just another example of the gap between the culture of science and the culture outside the lab. One should add that to her credit the director of the National Institute of Health intervened on that point, objecting that such a community standard "would have implicated even the discoverer of penicillin, who serendipitously found good use for bacteria growing in a contaminated lab dish" (as reported in the *Washington Post*, March 20, 1992).

The explicit or implied suspicion of fundamental corruption, of the basic flaw in the Ideal, when supported by the accounts of relatively rare but notorious cases and controversies, has so powerful an effect on the way science is presented and on the public debates, for three reasons. The first is that, with the growth in the size of the Federal financial support of research and development, there has been a rising discontent of the bureaucracy about regulating the enterprise and making it accountable. Second, ours is a time of unusual moral discontent when the public is staggering under wave after wave of accounts and charges of crookedness in business and government. But by themselves, neither source of discontent would explain the force of the assertion we are considering here. After all, those other betrayals of public trust, from Watergate to the financial scandals, are merely new evidence that Immanuel Kant was right when he wrote over two hundred years ago, "Out of the crooked timber of humanity no straight thing was ever made." Thus we really would be quite delighted if one day soon Congress abolished its sewer-money financing of campaigns, or if the whistle blowers in industry and government agencies would find they have to let their whistles rust in disuse. We would think it a premature April 1st joke if we were told that the American tobacco industry or the National Rifle Association were planning to hire this hall for an open meeting similar to this one, but on the ethics and values of their enterprises. Instead, the new moralists are more likely to focus on a group that, precisely because of its avowed incorporation of its own honor code, and because of its preference for organizing itself only poorly for public purposes, is singularly vulnerable to having guilt feelings induced in them: I speak of course of our research scientists. Among these, the rate of publication of provable fraudulent or falsified data has by one estimate been down at the astonishingly low level of around

0.002 percent,[2] owing chiefly to the internal mechanisms of validation — raising the interesting but neglected problems just why the rate of misconduct is so low, and how science on the whole can progress so well despite being done by fallible humans.

The enormous power of the generalized allegation against the conduct of science lies simply in the claim that basic research scientists in considerable numbers are false, and intentionally false, to their most fundamental avowed mission, i.e. to the pursuit of truths — the only legitimate pursuit their profession can still take pride in, since they gave up long ago portraying science in the service of religion, and tend to be embarrassed to draw their mandate and social tolerance for basic science chiefly from the spill-over into engineering applications. The accusation of fraud and of other misconduct, when used as a weapon against basic science, pierces a vital organ as it would in almost no other professions. By baring an absence of effective ethics at the heart of science, the place where the search for truth was supposed to be, the whole enterprise is delegitimated. It is as if it were discovered that a secret function of priests is to celebrate the Black Mass, or that the hoard of metal guarded in Fort Knox is mere tin.

Some might interject that in the long history of the shifting balance for the value of science to itself and society, no permanent damage has resulted from the dazzling variety of earlier charges made against it. Drawing almost at random from the bill of crimes posted just during the last one and a half centuries, science has been said to pervert the religious impulse; in the 1870s it was said to have lost all its ontological claims because it could not answer the fundamental questions about the nature of our consciousness and of matter, resulting in the famous cry "Ignorabimus" — we shall never know — by Emile Dubois-Reymond; following that, more than a decade was characterized by the fight over the "bankruptcy" of science; then Oswald Spengler's *Decline of the West* persuaded the masses that in the general metaphysical exhaustion of Western civilization, science was a cancerous cause; in the decades after World War I, the advances of science were widely seen to lead to technological unemployment; and in the last books of Lewis Mumford, science was accused of fostering a machine-like mentality responsible for over-optimistic and even inhumane uses of technology.

My favorite summary of the dark view of science and its disvalue is the anti-hero in Ivan Turgenev's novel, *Fathers and Sons* (1861). One of the greatest figures of Russian literature, together with Gogol, Dostoevski and Tolstoy, Turgenev was a poet largely in the tradition of 19th-century romanticism, inspired by Goethe, Schiller and Byron, among others. In *Fathers and Sons* the main figure is Yevgeny Vassilevich Bazarov, a university student of the natural

---

[2]According to the National Library of Medicine, during the period 1977-1986, 2,779,294 articles were published in the world's biomedical literature. The number of articles retracted because of fraud or falsification of data was 41—under 0.002 percent of the total. For further details, see Gerald Holton, *Thematic Origins of Scientific Thought: Kepler to Einstein*, rev. ed. (Cambridge, MA: Harvard University Press, 1988), pp. 457-58 and p. 470.

sciences, who expects to get his degree as a physician shortly. Being a scientist who "examines everything from a critical point of view," he confesses himself also to be a nihilist, the natural consequence of not acknowledging any authority. All talk of love and of the "mystic relationship between a man and a woman" is to him just "romanticism, humbug, rot, art." One should rather study the behavior of beetles. Reading the poetry of Pushkin, he says, is for little boys. He thinks it would be much better to start with a book such as Ludwig Büchner's *Force and Matter* (1855), which embodied such a flagrantly materialistic view that Büchner was forced to resign from his professorship in Germany. This, as it happens, is one of the books that most impressed Albert Einstein as a boy, and caused him to turn to the pursuit of science.

What matters, Bazarov says, "is that two and two are four — all the rest is nonsense." When he meets a clever and beautiful woman, he startles his friend by saying that hers would be a beautiful body to examine — on a dissection table. As if in revenge, fate brings him to the bedside of a villager dying of typhus, and he is made to help in the post mortem. But he cuts himself with his scalpel, and soon he is on the verge of delirium, a case of surgical poisoning. As he is dying, he tries to keep hold onto his kind of reality by asking himself aloud, "Now, what is 8 minus 10?" He is a caricature recognizable throughout literature—except that the figure of the emotionally dysfunctional scientist, from Dr. Frankenstein to the crew of Dr. Strangelove, cause surgical sepsis not only to themselves, but in all those around them.

As long as even such misguided scientists were not indicted as intentional crooks, there arose for every foolish Bazarov and Dr. Frankenstein the counter image of a noble, even heroic, Dr. Arrowsmith. At least in the public mind, all attacks sooner or later failed to stem the progressive incorporation of science into the central dynamics of our culture in recent times. And in fact the public continues so far to assert its uninformed but persistent approval. Scientists and policymakers take comfort from opinion polls that, over the past few decades, record favorable answers 70 percent or more of the time when people are asked a question such as "Do you feel that science and technology have changed life for the better?," or "Has science had a positive impact on the standard of living and working conditions?" The denial of the inherent credibility and value of science in modern society is not, or not yet, coming from the grassroots. It is coming from elsewhere, and it also makes use of a second element, added to the generalized charge of the rotten barrel rather than the occasional rotten apple. Let me now turn to that.

II

The enhanced and more sophisticated critique comes from an unorganized but fairly coherent assemblage, made up of members of a branch of contemporary philosophy of science and other humanists, the so-called "strong-program" constructivist portion of sociology and particularly of the sociology of science, a small subset of the media, a small but growing number of government officials, and a vocal segment of literary critics and political commentators associated with the so-called Postmodern Movement. It is a potent and eloquent collection of just the sort that in the past has successfully shaped and changed the worldview of their time and place.

The overall message that has been evolving over the past decade or two from that direction is a challenge no longer based only on the sensational stories of unacceptable behavior among scientists. The charge has been generalized and made more serious as follows: To put it in stark terms, the most basic fraud is one the scientific community has been committing all along, even those practitioners who did not intend any reprehensible breaking of the implicit honor code among them. For the basic fraud is the claim that there is any truth to be found at all. There is nothing there even to falsify; and conversely, science is inherently not corrigible, even if all misconduct were eliminated.

From that point of view, the business of science is mainly careerist, for example by keeping in operation expensive accelerators that claim to look for objectively ascertainable information about entities like quarks and bosons, which however are nothing more than socially constructed fictions. This is what the Cambridge University philosopher Mary Hesse meant by asserting that science can now claim no more than the status of a useful myth, and what the French sociologist of science Bruno Latour intended when he wrote we must "abolish the distinction between science and fiction." Others in this loose consortium of critics add that science has been a thinly disguised plot to maintain a largely patriarchal elite in power over the fate of minorities, women, and experimental animals.

The new definition such opinions lead to runs as follows: "science is politics by other means." The ancient rhetoric of the purity of science is a sham that must be exposed. When Vilfredo Pareto published his majestic *Treatise on General Sociology* (1916), he could still define science in those simple days as the "search for uniformities among facts independently of any consideration of utility, of sentiment, or of influence on conduct." In the middle decades of this century, scientists tended to adopt the more Baconian rhetoric that the acquisition of knowledge about the basic causes and interrelations of phenomena, by processes not easily predictable or fully understood, can yield the ability to exert power over those of nature's forces which cause our burdens and ills.

But now, the new consortium tells us, the arrow really goes the other way: not from knowledge to power, but from power to knowledge, and to a rather questionable knowledge at that. The old attempts to find generally applicable, shareable aspects of what might be called truth — through the use of the rational faculties of individuals, and through their skeptical but collaborative attempt to achieve some consensus — these attempts are not only doomed exercises, but ironically and perhaps inevitably they have led to the disasters that have marked this century. The whole modern era, launched under the flag of progress, has only led to tragedy. The extreme over-optimism of Herbert Spencer and Friedrich Engels can never be replaced by a soberer conception. Progress is illusion. Ours is the time to face what is now called the "End of the Modern Era" — the title of a powerful essay I shall come to shortly — and with it to the "End of Science" and the "End of Progress," titles of recent academic conferences held under distinguished auspices. Another new term for our present condition is that we are now in a state called the "objectivity crisis," a fashionable phrase which is also the title of recent conferences and of another significant document that we shall have to look at. Together, these aspects of the emerging sentiment spell the delegitimation of modern science as a valid intellectual force.   ·

I would like now to trace some of the individual steps and stages in this remarkable development of the growing new view, so as to make it easier to extrapolate and to preview the new terrain we may have before us. Here I can only point briefly to a few recent milestones on the road covered so far, and I shall do so by turning for help to recent writings by some of the most distinguished thinkers, rather than, say, through the Dionysian undercurrent of pop culture.

Our first informant and guide is Isaiah Berlin who to my mind is the most sensitive and humane historian of ideas. The collection of his essays, published recently as the fifth volume of his collected papers (Isaiah Berlin, *The Crooked Timber of Humanity, Chapters in the History of Ideas*, Vintage Books, a division of Random House, Inc., NY, 1992) opens with a startling dichotomy. He writes: "There are, in my view, two factors that, above all others, have shaped human history in this century. One is the development of the natural sciences and technology, certainly the greatest success story of our time — to this great and mounting attention has been paid from all quarters. The other, without doubt, consists of the great ideological storms that have altered the lives of virtually all mankind: the Russian revolution and its aftermath — totalitarian tyrannies of both right and left and the explosion of nationalism, racism and, in places, of religious bigotry, which interestingly enough, not one among the most perceptive social thinkers of the 19th century had ever predicted." He adds that if mankind survives, in two or three centuries' time these two phenomena will "be held to be the outstanding characteristics of our century, the most demanding of explanation and analysis."

What might Isaiah Berlin intend by so juxtaposing these two "great movements"? One's first temptation may be to chime in, as if by rote, with some standard responses to such a question. One of these might be that if it had not been for the ingenuity and frantic work of scientists among the Allies who produced radar, proximity fuses, and anti-submarine devices, supporting the valor of the allied soldiers in World War II, the totalitarian tyranny of that period might well have triumphed over the democracies, and established itself at least throughout Europe.

A second standard answer might be that these two "great movements" simply identify the polar opposites of mankind's persistent Manichaean tendencies. The great historian of science George Sarton put it in point blank terms:

"Definition: Science is systematized positive knowledge...."

"Theorem: The acquisition and systematization of positive knowledge are the only human activities which are truly cumulative and progressive."

"Corollary: The history of science is the only history which can illustrate the progress of mankind. In fact, progress has no definite and unquestionable meaning in fields other than the field of science." (George Sarton, *The Study of the History of Science* (1936: reprint New York: Dover, 1957, p. 5.)

Neither of these two responses, of course, would appeal to Isaiah Berlin. What is on his mind is quite different. As we follow his eloquent and subtle analysis, it dawns on the reader that science and tyranny, the two polar opposite movements which he holds to have defined and shaped the history of this century, are somehow connected — that the development of the natural sciences and technology in some sense may have been an unintended supporter, perhaps even a cause, of those "totalitarian tyrannies."

This stunning connection, to be sure, is never explicitly spelled out by him. But we can get a glimpse of the implicit argument later in the book, in his essay (published first in 1975) entitled "The Apotheosis of the Romantic Will: The Revolt against the Myth of an Ideal World." There, Isaiah Berlin summarizes the chronology of some familiar concepts and categories in the Western world, and specifically the changes in "secular values, ideals, goals." What commands his attention is the change away from the belief in the "central core of the intellectual tradition...since Plato," and toward a "deep and radical revolt against the central tradition of Western thought," a revolt which in recent times has been trying to wrench Western consciousness into a new path.

The central core of the old belief system that lasted into the 20th century rested on three dogmas that Isaiah Berlin summarized roughly as follows. The first is that "to all genuine questions there is one true answer, all others being false, and this applies equally to questions of conduct and feeling, to questions of theory and observation, to questions of value no less than to those of fact." The second dogma is that "The true answers to such questions are in principle knowable." And the third is that "These true answers cannot clash with one another." They cannot be incommensurate, but "must form a harmonious whole," the wholeness being assured by either the internal logic among or the complete compatibility of the elements.

It is out of these three ancient dogmas that, he says, institutionalized religions as well as the sciences developed their present form (although we might add that for about a century scientists, in their practice as well as in their philosophical discussions, have become more and more aware of the benefits of proceeding more non-dogmatically, by conjecture, test, refutation and probability). In their pure state, these systems are utopian in principle, for they are imbued by the optimistic belief, inherent in and derivable from the dogmas, that "a life formed according to the true answers would constitute the ideal society, the golden age." All utopias, Isaiah Berlin reminds us, are "based upon the discoverability and harmony of objectively true ends, true for all men, at all times and places" — and by implication the same is true for scientific and technical progress, which are aspects of our drive toward what he calls "a total solution: that in the fullness of time, whether by the will of God or by human effort, the reign of irrationality, injustice and misery will end; man will be liberated, and will no longer be the plaything of forces beyond his control [such as] savage nature...." This is the common ground shared by Epicurus and Marx, Bacon and Condorcet, the Communist Manifesto, the modern technocrats, and the "seekers after alternative societies."

But, Isaiah Berlin now explains, this prominent component of the modern world picture is precisely what the Romantic Rebellion, from its start in the German "Sturm und Drang" movement of the end of the 18th century, completely rejected, and swore to replace by the "enthronement of the will of individuals or classes, [with] the rejection of reason and order as being prison houses of the spirit." No one, he says, predicted that this worldwide growth would be what dominates "the last third of the 20th century." The Enlightenment's search for generalizability and rational order is depicted by the rebels of our time as leading at best to the pathetic Bazarovs of science, and those must be replaced by the celebration of the individual, by flamboyant antirationalism, by "resistance to external force, social or natural." In the words of Herder, the rebel shouts: "I am not here to think but to be, feel, live!"

This assertion of the will over reason has shaken or undermined what Isaiah Berlin had called the three pillars of the main Western tradition. The Romantic Rebellion arose, as it were, as an antithetical mirror image, created by the very existence of the earlier Enlightenment conception. It glows forth as its opposite, the "romantic self-assertion, nationalism, the worship of heroes and leaders, and in the end...Fascism and brutal irrationalism and the oppression of minorities." Moreover, in the absence of "objective rules," the new rules are those that the rebels make themselves: "Ends are not...objective values.... Ends are not discovered at all but made, not found but created."

In the end, "this war upon the objective world, upon the very notion of objectivity," launched by philosophers and through plays and novels, infected the modern worldview; the "romantics have dealt a fatal blow" to the earlier certainties, and have "permanently shaken the faith in universal, objective truth in matters of conduct" — and, he might have added, if he had then been aware of the new movements in philosophy and sociology, in objectivity in science itself. As with any revolt, we are confronted with mutually incompatible choices — either/or.

Other authors provide verification and elaboration of the implications of these findings. To glance at least at one telling example, the historian Fritz Stern has written about the early phases of growth of Nazism in Germany when there arose in the 1920s, in his words, the "cultural Luddites, who in their resentment of modernity sought to smash the whole machinery of culture." (I note in passing that similar analyses would result if we looked at the USSR or other totalitarian systems.) In Germany, the fury over an essential part of the program of modernity, "the growing power of liberalism and secularism," directed itself naturally also against science itself. Julius Langbehn was one of the most widely read German ideologues in the 1920s, and Stern writes of him "Hatred of science dominated all of Langbehn's thought.... To Langbehn, science signified positivism, rationalism, empiricism, mechanistic materialism, technology, skepticism, dogmatism, and specialization...."

Long before the Nazis assumed governmental power, some German scientists and other scholars demanded that a new science be created to take the place of the old one which they discredited — a new "Aryan science," based on intuitive concepts rather than those derived from theory; on the ether, the presumed residence of the "Geist"; on the refusal to accept formalistic or abstract conceptions, which were seen as the earmarks of "Jewish physics"; and on the adoption as far as possible of basic advances "made by Germans." In the careful study entitled *Scientists Under Hitler*, Alan Beyerchen identified some of the other main pillars of Aryan science, and it will pay us to listen for themes similar to those which are now also getting fashionable.

A prominent part of Aryan science was, of course, that science, as one would now say, is socially constructed, and therefore that the racial heritage of the observer "directly affected the perspective of his work." Scientists of undesirable races, therefore, could not qualify; rather, one had to listen only to those who are in harmony with the masses, the "Volk." Moreover, this *völkisch* outlook encouraged the use of nonexperts (provided they had the proper racial qualification) to participate in judgments on technical matters as lay persons, parallel to the expert (as in the *Volksgerichte*).

The international character of the consensus mechanism for finding agreement was also abhorrent to the Nazi ideologues. Mechanistic materialism, denounced as the foundation of Marxism, was to be purged from science, and physics was to be reinterpreted to be concerned not with the material but with the spirit. "The Aryan physics adherents thus ruled out objectivity and internationality in science....Objectivity in science was merely a slogan invented by professors to protect their interests." Hermann Rauschning, president of the Danzig Senate, quoted Hitler as follows:

> "We stand at the end of the Age of Reason.... A new era of the magical explanation of the world is rising, an explanation based on Will rather than knowledge. There is no truth, in either the moral or the scientific sense.... Science is a social phenomenon, and like all those, is limited by the usefulness or harm it causes. With the slogan of objective science the professoriat only wanted to free itself from the very necessary supervision by the State.
>
> "That which is called the crisis of science is nothing more than that the gentlemen are beginning to see on their own how they have gotten onto the wrong track with their objectivity and autonomy. A

simple question that precedes every scientific enterprise is: Who is
it who wants to know something, who is it who wants to orient
himself in the world around him?"[3]

On one issue, however, there was some division, namely, how technology, so useful to the state, could be fitted into the romantic ideal. In recent
times, many antimodern movements, including Fundamentalist ones, have
embraced technology. On the other hand, Philipp Lenard, an outstanding physicist although a chief cultural hero of Nazi propaganda, spoke for at least a
minority when he said that the tendency of scientific results to prepare the
ground for practical advances have led to a dangerous notion, that of man's
"mastery" of nature: Such an attitude, he held, only revealed the influence of
"spiritually impoverished grand technicians" and their "all-undermining alien
spirit." This idea, too, had its roots in the centuries-old history of the rise of
romantic thought in Germany. Alan Beyerchen summarizes this section with
the observation that "the romantic rejection of mechanistic materialism, rationalism, theory and abstraction, objectivity, and specialization had long been
linked with beliefs in an organic universe, with stress on mystery [and]
subjectivity...."

Some of these excesses — couched in words eerily reminiscent of those
we read in the current attempts to delegitimate the Enlightenment-derived
image of science and its values — would probably be seen by Isaiah Berlin as
coherent with and perhaps confirmatory of his own analysis of the Romantic
revolt. But as I now attend to the views expressed by another remarkable and
distinguished contemporary icon among humanists, we shall meet an advocate
rather than analyst. And we shall see that the relation between modern natural
science and the rise of totalitarianism, which Isaiah Berlin considered to be
only the result of an obscene abreaction, now receives a much more sinister
interpretation: the two become causally linked.

III

The point of view to which I am now turning to is embodied in a recent
address by the universally admired Czech poet, playwright, resistance fighter
and statesman, Václav Havel. You will notice that he embraces many of the
themes described in Isaiah Berlin's account; but Havel's key point is that totalitarianism in our time was simply the perverse extreme of a trend of ideas
embodied in the program of science itself. In this sense, Western science gave
birth to Communism; and with the fall of the latter the former has been irreme-
diably compromised.

Looking back on this century, other Central Europeans might characterize
it perhaps more as the playground of the forces of brutal irrationality and bes-
tiality, in which the fates of millions were sealed by the whims of Kaiser Wilhelm, Hitler and Stalin and their henchmen. But Václav Havel finds the chief
sources of trouble have been the very opposite, namely "rational, cognitive

---

[3]Hermann Rauschning, *Gespräche mit Hitler* (New York: Europa Verlag, 1940), p. 210.

thinking," "depersonalized objectivity," and "the cult of objectivity." I must let him put his case at some length in his own words, because he is a writer who eschews the documentation or balanced account of the scholar but who instead is in superb command of the rhetoric of persuasion, and of the chief art of the dramatist, the suspension of disbelief. The "end of Communism," he writes,[4]

"...has brought to an end not just to the 19th and 20th centuries, but to the modern age as a whole.

"The modern era has been dominated by the culminating belief, expressed in different forms, that the world — and Being as such — is a wholly knowable system governed by a finite number of universal laws that man can grasp and rationally direct for his own benefit. This era, beginning in the Renaissance and developing from the Enlightenment to socialism, from positivism to scientism, from the Industrial Revolution to the information revolution, was characterized by rapid advances in rational, cognitive thinking. This, in turn, gave rise to the proud belief that man, as the pinnacle of everything that exists, was capable of objectively describing, explaining and controlling everything that exists, and of possessing the one and only truth about the world. It was an era in which there was a cult of depersonalized objectivity, an era in which objective knowledge was amassed and technologically exploited, an era of belief in automatic progress brokered by the scientific method. It was an era of systems, institutions, mechanisms and statistical averages.... It was an era of ideologies, doctrines, interpretations of reality, an era in which the goal was to find a universal theory of the world, and thus a universal key to unlock its prosperity.

"Communism was the perverse extreme of this trend.... The fall of Communism can be regarded as a sign that modern thought — based on the premise that the world is objectively knowable, and that the knowledge so obtained can be absolutely generalized — has come to a final crisis. This era has created the first global, or planetary, technical civilization, but it has reached the limit of its potential, the point beyond which the abyss begins....

"Traditional science, with its usual coolness, can describe the different ways we might destroy ourselves, but it cannot offer us truly effective and practicable instructions on how to avert them."

The patient listener might at this point break forth with objections — that these passages are infested with illogical jumps and immense overgeneralizations, or that on factual grounds the self-designation of some Communist ideology as "scientific" was indeed a fraud, as demonstrated most simply in Lenin's own writings. But these would be dismissed as quibbles. The object of the piece is in its conclusion, in the "way out of the crisis of objectivism," as Havel labels it. Only a radical change in man's attitude toward the world will serve.

---

[4]"Politics and the World Itself," *Kettering Review* (Summer 1992), p. 9-11.

Instead of the generalizing and objectifying methods, we must turn, he says, to "such forces as a natural, unique and unrepeatable experience of the world, an elementary sense of justice, the ability to see things as others do...courage, compassion, and faith in the importance of particular measures that do not aspire to be a universal key to salvation.... We must see the pluralism of the world.... We must try harder to understand than to explain." Man needs "individual spirituality, firsthand personal insight into things...and above all trust in his own subjectivity as his principal link with the subjectivity of the world...."

Despite Havel's hint of a possible blending of the "construction of universal systemic solutions" or "scientific representation and analysis" with the authority of "personal experience," so as to achieve a "new, postmodern face" for politics, Havel's identification of the "End of the Modern Era" is not to be understood merely as a plea for some compromise or coexistence among the rival constructs; that much was announced in an earlier and even sharper version of his essay, one which dealt with the place of modern science quite unambiguously (Reprinted in Jan Vladislav, ed., *Václav Havel, or Living in the Truth*, London: Faber & Faber, 1987, pp. 138-139. The passage was written in 1984.):

"[Ours is] an epoch which denies the binding importance of personal experience — including the experience of mystery and of the absolute — and displaces the personally experienced absolute as the measure of the world with a new, man-made absolute, devoid of mystery, free of the 'whims' of subjectivity and, as such, impersonal and inhuman. It is the absolute of so-called objectivity: the objective, rational cognition of the scientific model of the world.

"Modern science, constructing its universally valid image of the world, thus crashes through the bounds of the natural world which it can understand only as a prison of prejudices from which we must break out into the light of objectively verified truth.... With that, of course, it abolishes as mere fiction even the innermost foundation of our natural world. It kills God and takes his place on the vacant throne, so that henceforth it would be science which would hold the order of being in its hand as its sole legitimate guardian and be the sole legitimate arbiter of all relevant truth. For after all, it is only science that rises above all individual subjective truths and replaces them with a superior, trans-subjective, trans-personal truth which is truly objective and universal.

"Modern rationalism and modern science, through the work of man that, as all human works, developed within our natural world, now systematically leave it behind, deny it, degrade and defame it — and, of course, at the same time colonize it."

Here we see the giant step which Václav Havel took beyond Isaiah Berlin's analysis: It is modern science itself that has been the fatal agent of the modern era; it has been responsible even for deicide.

IV

It is difficult to be unmoved by Havel's powerful mixture of poetical feeling, theatrical flourish, and bold evocation of ancient, bloodstained images. A summary of these ideas was published as Havel's OpEd under the title "The End of the Modern Era," on March 1, 1992 in the *New York Times*. It made immediate, convincing and uncritical contact with readers of the most varied backgrounds, including one especially well placed to ponder the values of science, and to draw conclusions of great import for the life of science in the United States. Here we have arrived at the last and most recent of the milestones on the road to the new understanding of the relation between science and values, and with it also to the third of the three elements pressing on the lever that determines the self-conception of contemporary culture.

The person who was so deeply affected by Havel's piece was none other than the distinguished chairman of the U.S. Congress Committee on Science, Space and Technology, a former physicist, and one of the staunchest advocates of science during his long tenure in the House, the Honorable George E. Brown, Jr., Democrat of California. As he explained recently ("Opening Remarks" for AAAS Panel; see below), he was inspired by Havel's essay in the OpEd version in the *New York Times* carefully to reconsider his role as a public advocate. He therefore had first written a long and introspective essay, published in September 1992 in the *American Journal of Physics* (v. 60, no. 9, pp. 779-81) under the title — borrowing from Havel — "The Objectivity Crisis," and then called together a group of social scientists in a public session at the annual meeting of the American Association for the Advancement of Science (February 12, 1993), under the title, "The Objectivity Crisis: Rethinking the Role of Science in Society."

These writings and comments show that Mr. Brown has clearly been persuaded by Havel's version of the Romantic Revolt, and has cast about earnestly for the consequence it should have for the pursuit of science in this country. As a pragmatic political leader, he is not primarily concerned with the question whether science has lost its epistemological and ontological warrant owing to the postmodernists' challenge to objectivity itself. Rather, he is necessarily concerned with how scientific activity may be legitimated by its service to society in terms of visible "sustainable advances in the quality of life," "the desire to achieve justice," which he says "is considered outside the realm of scientific considerations," and all the other "real, subjective problems that face mankind."

What is at stake, in his view, is the veracity of the Baconian promise — all that is left to science if one strips it of its claims to be honestly pursuing, or ever finding, truths. But even on that score, the remaining shreds of legitimacy are for him in doubt, just as we discovered earlier that the claims of truth-seeking, and of the evidence of truth itself, were denied. As George Brown puts it, he now sees little evidence that "objective scientific knowledge leads to subjective benefits for humanity." The implicit promise of progress must be doubted. The privileging of the claim of unfettered basic research is void too, he says, because all research choices are "contextual" and subject to the "momentum of history."

Moreover, science has usurped primacy "over other types of cognition and experience." Here, George Brown quotes Havel's definition of the "crisis of objectivity" being the result of subjugating our subjective humanity — our "sense of justice, ...archetypal wisdom, good taste, courage, compassion, and faith," to the processes of science that "not only cannot help us distinguish between good and bad, but strongly assert that its results are, and should be, value free." In sum, he says, it is all too easy to support more research, when the proper solution is instead "to change ourselves." Indeed, he concludes, "the promise of science may be at the root of our problems." To be sure, the energies of scientists may still find use if they are properly directed, chiefly into the field of education or into work toward "specific goals that define an overall context for research," such as population control — a narrow form of Baconianism.

When Mr. Brown presented his ideas recently at the AAAS meeting to the panel of social scientists whom he had selected, only one (Marvin Harris) disagreed openly and eloquently. But the general response was mild and acquiescent, and one of the other panelists revealingly urged Mr. Brown to go much further still, far beyond the widely agreed-upon mechanism for assuring the proper accountability of scientists in a democracy. He proposed that in arriving at future science policy decisions regarding major projects and controversies in science and technology, "we make strong attempts to involve ordinary citizens in processes of discussion and decision-making, including citizens who have not previously demonstrated expertise about such matters at all." And perhaps not realizing how close he was coming to the *völkische* solution tried earlier and elsewhere, including in Mao's "Cultural Revolution," he seriously suggested that our government form a variation of the National Science Foundation's Board, one whose membership would contain such nonexperts as "a homeless person [and] a member of an urban gang." No objections were raised by any of the participants. One felt as if one had a glimpse of a possible future.

In this brief overview, ranging from the trembling pillars of the Platonic tradition of the West to today's so-called end of the modern era, we have identified a historic trend that has risen and fallen and risen again. Today it represents only a minority view, but a view held in prominent circles, among persons who can indeed influence the direction of a cultural shift. If that trend should rise again to prominence, the new sensibility in the era to come will be very different indeed from the currently dominant one.

Of course, it may turn out that the present version of the Romantic Rebellion will peter out by itself — although I doubt it will. Or it may gain strength, as it did in 19th-century Germany and again prior to the totalitarian victories in the 20th. Or perhaps enough scientists, scholars and other intellectuals will cease their self-absorption, and will wake up to the need to rethink and assert their proper place in the culture of the future.

But such speculations, such doubts and hopes, are beyond the proper scope of the historian; and so I leave those in your hands.

Modern Day Hubris? Biotechnology and Genetic Engineering

Rita R. Colwell and Morris A. Levin

Introduction

Biotechnology is "technology," using Robert Malpas' definition: "Technology is the systematic harnessing of all knowledge and experience to produce something practical and commercially useful — a product, a manufacturing process, a system, a service, a methodology." Thus, technology is science plus engineering. Using the definition of Richard I. Mateles of Candida Corp., "Biotechnology is the set of concepts and laboratory or production scale techniques, which aim at the production of products through the growth, metabolism, or manipulation of cells or tissues of microbial, animal or plant origin, or of enzymes derived from them." This is, as Mateles says, a broad definition, but it excludes meat packing, farming *per se*, and medicine!

Until recombinant DNA (rDNA) technology, that is, genetic engineering, gene splicing, etc., burst on the scene, what we call biotechnology today was called biochemical engineering, fermentation technology, applied microbiology, industrial microbiology or similar titles. Obviously, the term "applied" was considered déclassé until the discovery that biotechnology allowed "real" money to be made. The term that was coined was genetic engineering — a set of concepts and techniques with wide application in biotechnology.

In 1984, the Office of Technology Assessment (OTA) of the United States Congress distinguished "old" biotechnology whereby beer, enzymes, antibiotics, solvents, and other products were made without rDNA, from the "new" biotechnology which *does* involve rDNA manipulation. The problem is, today, a decade later, rDNA methods are used to improve beer yeasts, wine fermentation, enzymes, etc.

Biotechnology, to draw from the vernacular, is the "new kid on the block." Acceptance of this technology, therefore, depends on, as with any "new kid," pre-existing feelings of the residents in the neighborhood. In the case of biotechnology, the feelings and belief systems of the residents are, in fact, those of the public. The "feelings of the residents," with respect to biotechnology, unfortunately, are influenced by past experience with "big science" and its linkages to commercial interests, i.e., big business. Thus, biotechnology finds itself in the position of being guilty until proven innocent. In analyzing this rather strange situation, the "kid," i.e., the biotechnology community, faces two major hurdles in addressing the problem of how to be accepted by the public.

The first relates to public opinion derived from adverse effects of relatively unfettered growth of science/industry linkages, the most dramatic examples derive from the era of chemistry ("better living through chemistry") and nuclear power (free energy). The public, unfortunately, has become immunized, as a result of adverse effects of these developments. One need only turn to events such as Chernoble and Three Mile Island to understand the situation. Toxic chemical sites, such as the Valley of the Drums and Times Beach, provide continuing flash-backs from the public's perspective.

Public perception of the outcome of science/industry relationships, in general, has been negative. Thus, biotechnology and the public are in a relationship that is based on pre-existing feelings of the residents that are very strong and hostile. The result is that the public is wary of connections that may develop in the future.

What are public opinions of biotechnology based on? How can they be modified? How can communication with the public best be achieved? Can these issues be addressed by providing credible, science-based risk assessment?

Another aspect of public acceptance of biotechnology involves more than quantitative, science-based, assessment of risk. The public, in fact, views risk differently than scientists. According to Morgan Granger (1992) of Carnegie Mellon, data from experimental psychology, using surveys of public opinion, have shown that when it is possible to assess the danger involved in an activity (in terms of hazard, when one equates hazard to death) the public can make this assessment, but may, at the same time, hold in its mind a different order of riskiness, i.e., activities to be avoided. To illustrate, if a person on the street is given a list of activities, he or she is able to rank them in terms of deaths per year. If asked to rank the same activities in terms of how "risky" they are, a different rank order will often be provided. This means, according to Morgan, that much more goes into evaluating risk than a specific hazard. Rearrangement takes these factors into account. Rearrangement is based on belief interacting with knowledge: beliefs or feelings are highly colored by age, sex, income, education level, and other factors.

An additional problem is the vision in the mind of the public of the mad scientist stereotype. He/she is in the laboratory obsessed with his/her scientific pursuit. As Hawthorne so eloquently states in "The Birthmark," the scientist returned to studies....

"in unwilling recognition of the truth — against which all seekers sooner or later stumble — that our great creative Mother, while she amuses us with apparently working in the sunshine, is yet severely careful to keep her own secrets, and in spite of apparent openness, shows us nothing but results. She permits us, indeed, to mar, but seldom to mend, and like a jealous patentee, on no account to make."

From this frame of reference, for the new kid, ethical issues are of much greater significance. Proving innocent and, therefore, acceptable involves more than just providing hazard analysis data indicating little risk. On the one hand, proving *no risk* is not possible, but on the other hand, no reasonable person would accept a claim of no risk. Risk alone is not the issue. The "risk issue," combined with public perception of science/industry connections, leaves us with uncertainty and ambivalence on the part of regulators, investors, funding sources, and the public.

The furor, raised when the first field tests of engineered microorganisms were proposed in the U.S. and when the industrial giant Hoechst proposed to open a biotechnology-based pharmaceutical plant in Switzerland, did not derive wholly from risk. Health and environmental risk issues were, indeed, raised, but these were eventually resolved when the data were made available. Yet, after all was said and done, the Swiss plant never opened. Despite formal approval in 1992, after almost three years of effort, the company decided to build the plant in another country. Field tests were conducted after a three-year delay. Paradoxically, since then, most countries have permitted field tests of genetically engineered organisms (GEO). Transgenic plants have been field-tested in 21 countries. Depending on how one does the counting, there have been *ca.* 1400 field tests of engineered organisms (GEOs). Without question, biotechnology is now a growth industry. The Food and Drug Administration has reported thousands of biotechnology products, either already approved for use or in the pipeline for near term approval (Figure 1). In the United States, more than 1500 companies list themselves as biotechnology companies. The majority were founded during the past several years, but a large number of older, well-established firms have joined the bandwagon and re-named themselves as biotechnology companies.

We can safely conclude that biotechnology is big business and will become even larger. June Grindley (1992) points out that "there is no doubt that biotechnology will have massive effects world wide throughout the major agricultural, food, drink and pharmaceutical sectors of industry." Others have pointed out that biotechnology will affect all industrial sectors. Through efforts of the UN, third world countries are actively developing biotechnology capabilities. Research in biotechnology has already produced results influencing all disciplines of the life sciences, and the effects will become more evident with time.

With the development of any big business there are societal, health, and environmental implications, each with beneficial and adverse aspects for both the long and short term.

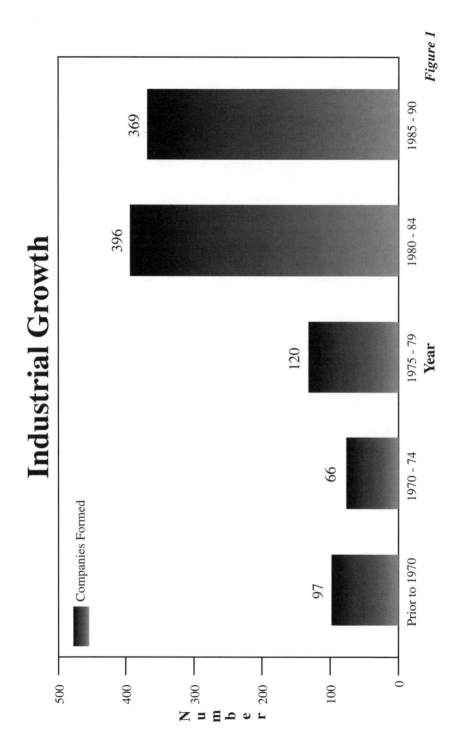

Figure 1

Some societal social impacts have already occurred. For example, university scientists, in general, were accepted by the public as being trustworthy. This is no longer the case, according to a University of North Carolina public opinion survey. The changed status is a result of biotechnology technology transfer. From the dizzying progress in molecular biology since 1990 and the almost insatiable industrial craving for new products, technology transfer in biotechnology is moving ahead rapidly. As more university/industry collaborations take place, more scientists are increasingly viewed as benefiting from the profits of industry and, hence, are not to be trusted as impartial observers any longer.

The relationships between public and private research, and the interactions within each, are shifting, essentially from basic to applied research, as a result of biotechnology technology transfer. Emphasis on application may have been a natural outcome, but biotechnology has certainly accelerated the process.

The resistance to biotechnology which manifested itself early on in the pyrotechnical development of biotechnology is still present and becoming even more apparent. The Hoechst plant situation and the bovine somatotropin (BST) are prime examples. As a result, biotechnology requires the appearance of being "squeaky clean."...it must become "ultracivilized" and prove (earn) its acceptability.

Having stated the above, nevertheless, many of the issues being raised concerning biotechnology genuinely are in the realm of the ethical or socioeconomic. Biotechnology must hew to ethical principles in a way no other area of science or industry has had to do before. To gain public acceptance and support, biotechnology must identify the ethical principles in which the public will believe and must demonstrate compliance, as well as convince observers that actions taken are not just because they are good for business or to ensure funding. To achieve this result, the principles of ethics must be identified, as well as actions taken showing that they are being taken seriously.

With this lengthy introduction, the focus of the remaining portion of this paper will be on: 1) identifying key ethical principles in biotechnology; 2) the need for this type of thinking in risk assessment; and 3) how and why basic ethical principles in biotechnology are related to successful commercialization and continued funding for biotechnology research.

From first principles, it is useful to examine selected reactions by some governments to biotechnology, including:

- reexamining how biological products have been regulated in the past. In the case of pesticides, this has resulted in a call from Congress to reregister all biopesticides.

- reviewing protocols to ensure that they remain applicable. The Food and Drug Administration has consistently held that the safety of biotechnology products can be assessed using existing protocols.

The National Agricultural Biotechnology Council (NABC), while not disagreeing with the content of the recently released regulations of FDA on labeling, wrote to all concerned that the process needs improvement, including the need to be more transparent, with more provision for public participation.

- reaffirming that the existing statutory framework was and still is adequate. This has resulted in the well publicized consolidated framework, the intent of which is to show that sufficient legislation exists to handle the new biotechnology products. The NABC report also points to a number of shortcomings or, at best, unresolved issues, in the regulatory framework, a few of which are as follows:

  - Should transgenic animals receive more extensive testing before slaughter for human consumption?

  - Should transgenic fish and shellfish and wildlife be covered by the regulations?

  - Should there be regulations for field testing of genetically engineered microorganisms (GEMs) in livestock feeds?

The NIH guidelines for work with transgenic animals have yet to be formally adopted; in any event, they are not binding on industry. The attempts to convince the public that biotechnology is safe and beneficial need to be articulated, pointing out that biotechnology provides needed products and services in a safe and cost effective manner, viz, humilin, biotreatment, and biopesticides. For example, the USDA developed an exhibit, which premiered at the Smithsonian and, subsequently, was taken to other locations. The exhibit describes the benefits and safety of biotechnology.

It is anomalous that the government is not looking into why negative pressure against biotechnology exists, or developing methods to counteract it. The Office of Technology Assessment of the United States Congress produced a series of reports discussing public involvement with biotechnology. The surveys focused on education issues, acceptance of risk, and safety from a personal, physical or environmental perspective, but little attention was given to socio-economic or ethical issues.

The first four points above concern safety, in the scientific sense, albeit mixed with politics in the area of adopting regulations. The last point is germane to the issue of ethics/societal issues and acceptance. There may be issues, below the surface, yet to emerge. It is possible, for example, that the real basis for opposition to biotechnology is fear of change, fear of altering well established norms. It is obvious that the rate of change has accelerated in recent years. Fear of further changes, a desire to keep things the way they are, i.e., familiar and friendly, may be a major factor in the opposition to biotechnology. Nothing epitomizes this fear more clearly than the following letter to President Andrew Jackson from Governor Van Buren dated January 31, 1829:

To President Jackson:

The canal system of this country is being threatened by the spread of a new form of transportation known as "railroads." The federal government must preserve the canals for the following reasons:

One: If canal boats are supplanted by "railroads," serious unemployment will result. Captains, cooks, drivers, hostlers, repairmen and lock tenders will be left without means of livelihood, not to mention the numerous farmers now employed in growing hay for the horses.

Two: Boat builders would suffer and towline, whip, and harness makers would be left destitute.

Three: Canal boats are absolutely essential to the defense of the United States. In the event of the expected trouble with England, the Erie Canal would be the only means by which we could ever move the supplies so vital to waging modern war.

As you may well know, Mr. President, "railroad" carriages are pulled at the enormous speed of 15 miles per hour by "engines" which, in addition to endangering life and limb of passengers, roar and snort their way through the countryside, setting fire to crops, scaring the livestock, and frightening women and children. The Ambience certainly never intended that people should travel at such breakneck speed.

> Martin Van Buren
> Governor of New York

Dr. Jefferey Burkhardt of the University of Florida expressed the opinion that there are three major ethical problem areas: the lack of articulated professional ethics; the changing mission of public institutions; and our collective responsibility toward future generations for the environment as well as the future of science.

Governments, and many scientists, have, by and large, ignored the ethical issues. For example, Miller et al (1990) writing in *Science*, discuss risk-based oversight of experiments carried out in the environment and examine in depth whether evaluation of biotechnology products should be based on the process by which these are produced or solely on the product itself. The impact of the product on the public, or on the producers, e.g. the farmers, or on the biotreatment industry is not mentioned.

Selected Ethical Issues

Standard risk assessment of biotechnology products comprises the estimation of potential adverse effects in terms of damage to public health or to the environment. In some cases, risks are weighed against benefits before approval of a product or a field test is granted. The idea of including ethical or moral issues in the risk assessment process has created dissension among evaluators and the industry. It is difficult to establish endpoints for risk assessment — even in relatively simple situations dealing with a specific material and a limited population. There are serious problems in determining an appropriate end point for environmental applications. Imagine a regulator faced with determining a sociological or ethical endpoint for quantitative risk assessment. The public, or at least the interest groups which claim to represent the public, suggest that, in order to gain public acceptance of a product and public confidence in biotechnology, such issues *must* be considered.

Societal reaction is predictable. Pressure for regulations (to provide some form of protection) comes from the public directly or via interest groups. Concerns are mostly related to actual, measurable, physical effects on health and environment. As stated earlier, biotechnology is adversely affected by prior bad press of other science/industry-related issues. However, the cause of the resistance may be ethical and pocketbook issues. That is, the small farmer fears engineered seeds may become too expensive or the overall yield of his crop will lower, despite decreased loss to insect pests, and the loss may force him out of business (the cost of additional fertilizer to overcome the yield loss). Yet, the product appears to be safe from both the health and environmental perspectives. The unanswered ethical and social questions form a nucleus for accumulating resistance to what is perceived as a threat to one's livelihood or way of life. Grindley (1992) points out that protagonists for biotechnology products have mentioned, up to now, that silence is golden. They believe that, because of the beneficial aspects of products (which have been assessed for safety by appropriate government agencies), the products will sell themselves.

To understand these issues one must look at the relationship between socio-economic evolution and genetic resources, as Calestous Juma (1989) offers in, *The Gene Hunters*. He points out that Thomas Jefferson once said that the greatest service that could be rendered to any country was to add a useful plant to its culture. He was, of course, talking about the United States in the era in which he lived. Juma's point is that economic history often focuses on technology and ignores the role of genetic material.

Historically, there have been three major phases in the evolution of biotechnology. The first generation of biotechnologies (dating as far back as 7000 BC) derived from empirical practice. The products of fermentation were food (bread) and drink (wine). Knowledge was communal and passed on to succeeding generations as part of the cultural heritage. This very long period of time can be viewed as lasting until the Industrial Revolution. After the Industrial Revolution, a combination of advances in biochemistry and engineering led to improvements in fermentation processes and yielded new processes.

The result was an abundance of products and the beginning of the conquest of many diseases. According to Juma, "the history of industrialization is partly a story of the separation of production knowledge from its end use products." New products had been the result of slow development, allowing ample time for people to become accustomed to them and to see and weigh their value and overall effect. Mistakes were important locally but not nationally or globally. In terms of perception of biotechnology, one must remember that in the "Age of Chemistry," the value of the chemical industry was in the millions of dollars. The value of nuclear industry was measured in the hundreds of millions, but, in contrast, biotechnology is in the billion dollar range.

Major scientific advances, such as those by Watson/Crick and Cohen/ Boyer, led to revolutionary developments in biology, notably in molecular biology and ultimately in biotechnology. The Nobel Prize-winning research resulted in the opening of the third and present phase in biotechnology. The pace of change, indeed, intensified. For thousands of years progress was very slow, almost impossible. Over the past decades there were few major changes, leaving little time for adaptation. We are now faced with rapid change and with the need to evaluate products and processes before we see them in action.

Gary Comstock (1989 a) draws clear distinctions between types of issues. He points out that reservations or concerns relating to genetic crosses over species or kingdom boundaries, the impact of the product on small farmers, the power of large chemical/agricultural companies, or problems associated with developing a monoculture based agricultural sector are moral problems. These issues cannot be resolved by scientific analysis of experimental data. Therefore, input from the world of ethics is required.

The law of gravity describes the falling of petals from dying plants. The petals fall because of physical attraction. Science does not pretend to tell the petals what to do next. Science describes.

Ethics prescribes. Whereas science explains and predicts, ethics tells us not what is done but what ought to be done. It must be pointed out that simply because something happens (or does not happen) in nature does not mean we are morally unjustified in altering it. Philosophers refer to this as the is/ought fallacy. Comstock's example is that "just because the slave owners coerced their slaves into picking cotton does not mean that people ought to be enslaved."

Where scientists ask "What is going on — What can be done" philosophers ask "What ought to be going on." Answering the ethical questions requires both the best available data and the best available humanistic reflection and philosophical theory.

A group of environmental professionals, the American Institute of Certified Planners (AICP), developed a code of ethics for its members. The code contains two statements:

1. Recognize and attempt to reconcile societal and individual human needs with responsibility for physical, natural and cultural systems.

2. Promote and develop policies, plans, activities, and projects that achieve mutual support between natural and man-made, and present and future components of the physical, natural and cultural environment.

Examples of Specific Problems

Given the need for establishing the ethical dimensions and, in light of the AICP code, examples focusing on public health (GEOs in bioremediation) and environment (GEOs in agriculture), are most appropriate to consider, in terms of how to proceed with risk assessment and, perhaps, product or research selection.

Bioremediation

Bioremediation risk assessment is highly complex because of the wide range of issues and the difficulty in obtaining precise field data. Large tracts of land, entire aquifers, and specific dump sites are so heavily contaminated that human health (of those living nearby and using the groundwater) and environmental effects are evident. The ultimate cure is obvious to all. Rid the area of the pollutants. The mechanics are not obvious. Use of microorganisms is increasing, but there is much uncertainty as to the use of GEMs. Other candidate protocols, such as burying, fixing, and incinerating are available, but each has its own drawback. None of the three actually destroys (renders totally harmless) all of the polluting material. The material is merely changed in form (to glass bricks or incinerator ash). For example, incineration and air-stripping have advantages of decreasing the amount of toxic material. However, incineration is energy-demanding and may also lead to production of additional toxic materials, e.g., dioxin, which requires costly scrubbing before release into the atmosphere. Air-stripping also results in small amounts of material released to the environment. In addition, depending on the method used, it may produce the additional problem of transferring toxic materials onto activated carbon, which will then require treatment. In many cases, combinations of biological and physical/chemical treatment are preferable because of economic and time considerations.

Bioremediation is one of the fastest growing areas of biotechnology. Microorganisms capable of carrying out degradation process are varied. In some cases, they have been identified and characterized, whereas in others, even growth of the culture in isolation has proved to be extremely difficult. One of the best studied organisms is *Pseudomonas* G4, a bacterium effective in degradation of TCE. The organism is able to be induced when placed in the environment in the presence of a large number of co-substrates. Because it can degrade several compounds, as well as be induced by other compounds, G4 may have great utility in remediation of wastes from sites where a mixture of toxics is present. Genetic modification of this bacterium should enhance its abilities significantly.

Genetic modifications of microorganisms for the purpose of bioremediation have been widely investigated. Those modifications yielding an increase in the range of capabilities of microorganisms, as well as the rate of degradation and stability under environmental stresses have been attempted. While excellent results have been obtained in laboratory situations, scale-up studies are still in the developmental stage. Land-farming is perhaps the simplest and most common method of treating wastes, because it is done directly on site and simply involves addition of nutrients and moisture, as required, to achieve microbial growth on, and metabolism of, pollutants. Large amounts of soil can be treated, requiring only minimal effort. The physical composition of the site containing the hazardous material will have a significant effect on the ability of the microorganisms to carry out remediation. It is important to determine the hydrogeology, location of underground pipes, etc., which will affect the ability to expose added nutrients and oxygen directly to the bacteria carrying out the degradation processes.

Problems associated with fluctuations in environmental conditions include partial degradation, resulting in the presence of intermediate metabolites. For example, the pathway for the degradation of tetrachloroethylene (PCE; a known animal carcinogen) can result in the accumulation of vinyl chloride, which is also a human carcinogen. McCall et al. (1981) reported that, during the degradation of PCE, concentrations of 2,4,5 trichlorophenol and 2,4,5 trichloroanisole were found in the soil.

Problems associated with added nutrients may also exist. Soil contains large numbers of many types of microorganisms, including plant, animal and human pathogens. Any of these can benefit from added nutrients, in terms of enhanced survival or growth.

Currently, decision-making as to whether to permit a test or product deployment involves estimating health and environmental risks. Relative to the AICP code and the ethical issues described above, a policy is needed that asks "what ought to be done? Is the disease worse than the potential effects of the cure?" There have been no adverse effects from biotreatment recorded to date, but there have been no field applications of engineered organisms to date. All applicators have used existing, i.e., *in situ* and naturally occurring (autochthonous) microorganisms. Research in the development of GEMs is ongoing but slow, because of lack of a clear policy on how to evaluate GEMs for biotreatment purposes.

Questions which would — should — be raised in the development of such a code include:

- Are adverse health or environmental effects probable as a result of adding nutrients or incomplete digestion? — probably not (none noted to date), but oversight and more data are needed.

- Are adverse health or environmental effects probable as a result of allowing the pollution to persist as a result of treating the area by solidification or incineration? Probably, and these are quantifiable.

- Is it appropriate (ethical) to go ahead with biotreatment before being absolutely certain of no adverse effect? How does one decide?

The framework for such input — a code of ethics to be adhered to — must be developed. Consideration of the ethical perspective can alter the final decision in either direction. For example, one possibility is to look at the greatest good for the greatest number — and count future generations in the estimate. In the case of ground water contamination, effects can be very long term. The effects of "blooms" of unwanted microorganisms or intermediary metabolites could be measured over much shorter time frames. Input from those directly affected, i.e., the immediate public, and from the view of organizations/groups/agencies responsible or charged with consideration of the rights and needs of society at large are essential.

Agriculture

There is no doubt that biotechnology research has resulted in and promises to produce major advancements in agriculture. Comstock (1989 b, 1991) describes benefits, such as decreased herbicide use, the advent of degradable, specific biopesticides, and some of the potential adverse effects from a biology perspective (evolutionary, impact on gene pool, gene transfer, community effects) and from a social perspective.

It must be stressed that no adverse effects have been noted to date. Companies have applied for patents on specific animals (Merck, Macro Chicken, European patent on a chicken with the bovine growth gene). Australian workers are developing sheep which secrete insect repellent into their hair follicles, which will provide resistance to the blowfly (and also yield moth resistant wool). Cattle resistant to specific diseases would expand the area available for cattle farming (especially in Africa). Animals (and plant cells) which can produce pharmacological products i.e., bioreactors are being developed.

Yet, there is mistrust and lack of acceptance of biotechnology. If cattle raising areas are expanded, what happens to the native wildlife? Are the MacroChickens highly susceptible to particular diseases, i.e., could we have a repeat of the corn blight? Will the cost of the eggs or breeders drive small producers out of business?

Many of these questions could be or have been raised and/or resolved in advance. Communication of the resolutions to the public and to interested parties would diminish concern. NABC 4 was held in College Station Texas, starting on May 25, 1992. One day before the meeting, the FDA announced its policy for evaluating the safety of foods of transgenic plant origin. Several groups at the meeting coalesced to convince the organizers to write to FDA, the

Vice President's Council (no longer in existence), HHS, EPA, and USDA representatives to express NABC's belief that acceptance by the public can be expected only when opportunities for interaction are present. Policies should be developed in a transparent manner, with appropriate input from all concerned parties. Development of policy making as a process involving all parties on a partnership basis is essential to assure development and acceptance of biotechnology research and subsequent products.

The present system lacks features conducive to generating trust on the part of the public. As pointed out there are gaps in the regulations. In some cases, guidelines exist but have not been formally ratified (NIH Q for animals, P for plants and Greenhouses). In some areas, no oversight mechanism exists. There is no oversight mechanism for commercial scale release of transgenic animals (or of second generations of transgenic animals). NABC 4 identified a series of gaps (9) and additional issues best termed as "Should." These include:

- Should the consumer be able to make a choice via labeling with respect to using TG products?

- Should TG animals used as reactors receive special attention?

- Should there be a greater role for individual states?

Biotechnology is a complex science. Regulation is even more complex, partly because of the topic and partly because of the presence of past, somewhat but not completely applicable regulations and statutes, and natural reluctance to develop new ones. Many groups have stressed the need for a better educated public. Many groups have stressed the need for more openness on the part of regulators.

Development of a set of ethical principles is a basic need which has largely been ignored. The principles would provide guidance to regulators, interest groups and interested parties, as to what ought to be. The NABC, following "intense discussion on concerns," identified eight possible harms or benefits (Figures 2 and 3) which provide a possible starting point. Note that of these four benefits, one could argue that most have a moral element, requiring ethical input for decision making. Thus, removal of genetic defects or a better understanding of the well-being of cattle could result in fitness for new areas, with the concomitant problem of displacement of existing populations. Development of more efficient production raises issues of animal well-being.

The four possible harms all have clear moral implications. Animal rights, well-being, and impact on traditional husbandry stand out.

Medical Biotechnology

The ethical issues related to medical biotechnology have been raised and discussed during the past 15 years. Issues related to somatic versus germ line therapy and questions related to the use of medically oriented information

Possible Harm from AgBiotech

1. Diverting resources from traditional husbandry

2. Loss of genetic diversity

3. Development of genetically defective animals who suffer from disease as models for studies of disease

4. Thinking of animals as human artifacts

Figure 2

Possible Benefits from AgBiotech

1. Removal of genetic defects from animal population more rapidly

2. Better understanding of animal well being

3. Increased disease resistance

4. More efficient production (resulting in use of fewer animals)

Figure 3

from research on the human genome and, more recently, from the Human Genome Project are still being discussed.

From a strictly medical perspective, David Danks (Chairman of the Human Genetics Society of Australia) points out (Danks 1992) in a review of the Australian Parliamentary report on development, use, and release of engineered organisms that there is nothing new, and from his viewpoint, no real problem. He also distinguishes clearly between therapy and research, pointing out that research is needed to determine what is possible, to provide more options and extend knowledge. The report contained 48 recommendations, only two of which related to gene therapy and medical biotechnology. Danks states that these had been raised and discussed more than a decade ago, and that the recommendations made then can be supported now. These include the continuation of gene therapy at the somatic level and the need for a separate inquiry on heritable therapy.

Somatic therapy is to be conducted under the guidelines of the (Australian) National Health and Medical Research Council. The evaluation conducted is to assure the safety of each therapy (via animal tests) before being applied to humans and "to make certain that the inserted gene will be confined to the target tissue and not enter the germ line." A secondary concern is that the therapies should be used only when the disease places an "extreme burden" on the patient.

The recommendation relevant to germ line therapy called for an additional board of inquiry. Danks' position is that there is no need for this type of therapy and that the board of inquiry would have little to discuss. He believes that, since it would be necessary to identify defective fertilized eggs at an early stage and then alter the genetic makeup, it would be wiser to simply implant an unaffected egg or embryo. This obviates the need for treatment. In addition, by using this approach the effect does not carry through to succeeding generations. Note that this assumes development of technique to evaluate fertile eggs at an early stage.

Danks does point out that the use of germ line therapy to improve normal characteristics would be possible and that this would be a matter requiring major ethical discussion.

Relative to the information explosion, Juengst (chairman of the Ethical, Legal and Social Implications Program of the Human Genome Project) and Watson (director of the project) believe that "as biomedical sciences mature, biomedical scientists' social responsibilities also grow." They (Jeungst and Watson 1992) point out that:

- Human genome research will greatly increase the number of diagnostic and prognostic tests available

- There will be an increased risk of misinterpretation, with resulting possibility of trauma, stigmatization, and discrimination

- Information from the tests is of the most personal nature and most wish it to remain confidential

- These facts lead to the conclusion that the social responsibilities of biomedical scientists will increase.

There are three major sets of questions which have been identified through the efforts of the Human Genome Project. These include integration of the new tests into health care, educating and counseling the public, and problems of limiting access to and use of information by third parties. These are interrelated, major problems.

Integration of tests into the existing health care infrastructure raises questions of accuracy and quality control, defining the indications calling for the particular test, design of test protocols, training the physician and counsellor, controlling access to the data, and reimbursement. In addition, the public — the user — must be better educated to understand the evaluative assumptions prevalent in interpreting the findings. This will become more important as more tests designed to detect susceptibility to particular health problems come on line. These issues are knotty and difficult and, while some are more amenable to hard data gathering (equality control), all have an ethical aspect, e.g., how accurate must a test be before it is permitted on a mass basis?

Use of data by third parties presents an even greater ethical challenge. Clearly, maintaining confidentiality, though difficult, is possible. However, genetic information almost always has implications for other people's welfare as well as that of the patient. On a broad or societal level, insurers and employers may be affected. On the personal level, spouses and children are, or could be, directly affected.

Each of these questions raise doubts about the acceptability of the products of biotechnology, if not of biotechnology itself. Clearly, standards are required to aid in the evaluation of the technology and its fruits.

References

American Institute of Certified Planners. 1981. Code of ethics and professional conduct. The institute of certified planners, Washington, DC.

Cantely, M. 1987. Democracy and Biotechnology. Popular attitudes, information, trust and public interest. Swiss biotech. 5(5) 177-184.

Comstock, G. 1989 a. Genetically engineered herbicide resistance. Part one. *J. Agr. Ethics.* 2:263-306.

Comstock, G. 1989 b. Genetically engineered herbicide resistance. Part two. *J. Agr. Ethics,* 3:173-210.

Comstock, G. 1991. Is genetically engineered herbicide resistance compatible with low input sustainable agriculture (LISA)? NABC 1, 1989 NABC, Boyce Thompson Inst. Ithaca NY 14853.

Danks, D.M. 1992. The human gene therapy response. *Australasian Biot,* 2: (3) 159.

Granger, M. 1992. Academic controversy on risk in: *Center for Biotechnology Policy and Ethics Newsletter,* Texas A&M, 2(1):2.

Grindley, J. and D. Bennet. 1992. Public perception and the socioeconomic integration of biotechnology. *UK Biotechnology Handbook* 1992.

Jeungst, E. and Watson, J. 1992. Human Genome research and the responsible use of new genetic knowledge. *The Genetic Resource* 2(6):41-44.

Juma, C. 1989. *The Gene Hunters.* Princeton Univ. Press, Princeton, NJ.

Levidow, S. and J. Tait. 1991. The greening of biotechnology: GMOs as environment friendly products. *Science and Public Policy. 18*(5):271-280

McCall, P., Vrona, S. and Kelly, S. 1981. Fate of Uniformly Carbon-14 Ring Labeled 2,4,5 Trichlorphenoxyacetic Acid and 2,4 Dichlorophenoxyacetic Acid. *Journal of Ag. and Food Chem., 29*:100-107.

Miller, H., R. H. Burris, A. K. Vidaver and N.A. Wivel. 1990. Risk Based Oversight of Experiments in the environment. *Science, 250*:490-491.

National Association of Environmental Professionals. Code of ethics and standards of practice for environmental professionals. *Env. Prof. 9*(3):1987.

Sagoff, M. 1989. Biotechnology and the end of medicine. *Phil. and Med. 88* (2):85-87.

UK Royal Society Report. Sir William Bodmer. 1985 Royal Society, London.

Skepticism About Teaching Ethics

Bernard Lo

Teaching about ethical issues in scientific research constitutes a growth industry. Professional journals such as *Science* have published prominent articles on error, misconduct, and conflicts of interest in science. *Time* and *Newsweek* have featured these issues in cover stories. Many universities have established courses and symposia to discuss these topics. All of us attending this Sigma Xi symposium believe these ethical issues are important and deserving of attention. But many of our colleagues would disagree, particularly those in the target audience. Senior scientists may regard discussions of ethics as unnecessary, unhelpful or even counterproductive. Many young scientists attend such courses and seminars on ethical issues in science only because they are required to do so, often as a condition of their training grant. They would rather be in the laboratory, carrying out their research. We need to try to understand the reasons for this skepticism or opposition to teaching about ethical issues in science. This paper will discuss several objections that skeptical scientists commonly raise and make some suggestions for helping students learn about ethical issues.

Reasons for Skepticism About Teaching Ethics

Some scientists believe that teaching about ethics is ineffective because doing right and wrong is a matter of character. Others believe that studying ethics is unnecessary because they already know how to deal with ethical issues. Still other scientists acknowledge that they would like help with ethical issues but doubt that studying ethics will help them. These objections to teaching about ethical issues in science deserve closer attention.

"Only unethical persons have ethical problems."

Skeptic: "Unethical behavior is flagrant, such as making up data for experiments which never were performed. Of course there are some corrupt people in scientific research, as in any field. But it insults honest researchers to suggest that the everyday practice of science presents ethical issues."

Response. This is the bad-apple theory of misconduct: ethical problems are caused by a few morally defective individuals, not by the environment in which science is practiced or even the nature of the scientific enterprise. By implication, if the bad apples can be identified, ethical issues would disappear. If you are a good apple, ethical issues need not be a major concern.

Of course, clear-cut fraud is wrong, and people who do it deserve moral censure. Studying ethical issues will never guarantee that such people will do what is right. But ethics also deals with situations in which there are strong reasons both for and against a course of action. Decisions may be difficult because ethical guidelines conflict, or their interpretation is unclear. For example, it is not always clear how to distinguish between creative interpretation of data, honest error, and intentional fraud, as has been cogently discussed by other speakers with regard to Millikan's experiments on the charge of the electron. To acknowledge that such ethical issues are difficult shows realism and courage, not moral failing.

"Ethics is being a good person, not a system of rules."
Skeptic: "The best way to resolve ethical dilemmas is to have a good scientist whom you trust to do the right thing. Rules don't help that much. At some point, rules need to he interpreted. Then it's crucial to have an honest, thoughtful person on the case. Furthermore, there are people who follow the rules literally, but only because they want to avoid legal problems. You wouldn't hold them up as models for young scientists."

Response: Good scientists certainly act in praiseworthy ways that could not be readily codified in rules, such as rules about authorship of publications or retention of primary data from experiments. But being a good person — a virtuous person in the language of philosophers — is neither necessary nor sufficient for appropriate action. Rules and guidelines are helpful as well. First, scientists who lack virtue may still act correctly. Selfish, manipulative people may still observe the rules for authorship or retention of primary data. Second, good scientists may be genuinely perplexed in a given situation. For example, a thoughtful scientist may not know how to respond to an allegation that one of her postdoctoral fellows has misrepresented the data. Ethical guidelines set explicit expectations for conduct. Third, scientists are human. They make mistakes and work under stress. It would be inspiring if scientists always acted in an exemplary manner, but it would be unwise to assume they will. Ethical rules are like insurance, protecting against scientists whose character is flawed, who suffer lapses of character, or who are overwhelmed by a difficult situation.

"By the time you're a scientist, your ethics are set."
Skeptic: "You learn ethics as a child, in your family, church, or synagogue. By the time you're a scientist, your ethics are already set."

Response: While people learn basic moral precepts as children, scientists face many ethical issues for the first time during their training. For example, laypeople have no experience with issues of authorship of publications and management of research data. Thus while studying ethical issues does not guarantee that scientists will act appropriately, it helps them think through unfamiliar issues.

"Ethics is common sense and experience."

Skeptic: "Resolving ethical issues is just a matter of common sense and experience. You don't learn it from books. You learn it from day-to-day work in the lab."

Response: Common sense and experience are always helpful. Many ethical dilemmas, however, are so complicated that sensible and experienced scientists may be perplexed or may disagree over what to do in a particular situation. For example, scientists may disagree over how to interpret discrepant research data. Even when scientists intuitively understand the right action in a particular situation, it is helpful to make their thinking explicit. By providing an approach and guidelines, ethics can help scientists explain their reasoning to students, colleagues, and the public.

"Every case is unique, so guidelines are impossible."

Skeptic: "Every case is unique. You have to decide each one individually. You can't just follow guidelines."

Response: To be sure, general guidelines cannot be applied mechanically. They must he interpreted in particular cases, and there are always justified exceptions to guidelines. Practical wisdom, discretion, and judgment are always needed. On the other hand, ethical guidelines can be developed, even though cases are so individualized. Certain paradigmatic situations and dilemmas recur, such as problems with missing primary data. Cases that present similar ethical problems under comparable circumstances should be resolved in consistent ways. By thinking through decisions in a range of related cases, scientists can identify and formulate general guidelines that can be useful in future cases.

"Ethics is merely personal belief."

Skeptic: "In the tough cases I face, there are no right answers. There are good arguments for and against everything. It comes down to personal beliefs, and one person's opinion seems just as good as another's."

Response: Some arguments are more convincing than others. Explanations may be flawed because they are internally inconsistent, or do not respond to countervailing arguments. Everyday experience tells us that people with strong beliefs can be persuaded by cogent arguments and that people with widely different world views can agree in specific cases. In addition, individuals often hold several fundamental beliefs that are in conflict. For example, a scientist whose fellow is accused of misrepresenting data may believe both in trusting her colleagues and requiring rigorous data. Studying ethics can help people reconcile conflicting beliefs by clarifying the arguments that justify them.

"Studying ethics doesn't help solve real problems."

Skeptic: "Ethics is too abstract. Books make it all sound so logical, but it's not that way with real life problems. I don't have the luxury of thinking about one ethical issue at a time. I have to deal with several issues at once. I have a lot of people very worked up about problems. I need specific answers."

Response: The author agrees completely that studying ethics should help scientists facing real-life ethical dilemmas. Let us now discuss several suggestions for doing so.

Suggestions for Teaching About Ethical Issues

The title of this panel is *teaching* ethics and values. It may be helpful, however, to reframe the topic as *learning* about ethics. Learning focuses attention not on teachers or curricula but on those who learn, whether they are students or more senior scientists. Teachers need to start with the questions and assumptions that the students bring to the subject. Discussions therefore need to take into account the objections discussed previously.

Generally students learn best when their interest is captured and when they participate actively in learning. This may be particularly true regarding ethical issues, which by their nature may involve strong beliefs. Students usually learn more when they think the issues through for themselves than when they listen passively to lectures on ethical issues. This can be accomplished in several ways.

Discuss realistic cases. Cases can capture the students' attention and make issues come alive. The cases discussed should ring true to scientists and students. They should not be contrived "thought exercises" that make a philosophical point but would never occur in reality. Let me give two examples of cases that can stimulate discussion.

Case 1. *Interdisciplinary research*. You have purified a protein that you believe is a neurotransmitter receptor. For a functional assay, you collaborate with an electrophysiologist. From patch clamp recordings of the receptor, the electrophysiologist concludes that the protein does function as a channel specific for a particular ion. The two of you start writing the paper. Do you review the primary data from the electrophysiology experiments? Do you review the recordings with the electrophysiologist to select the "example of a typical recording" for publication? Are you responsible and accountable for the electrophysiology part of the paper?

Case 2. *Problems replicating results*. A new graduate student joins your lab intending to continue the research of a recently graduated student. The new student has mastered the experimental technique but cannot replicate the results of the departed graduate student. What do you do as the head of the laboratory if the graduate student comes to you with this problem? What if the new student not only cannot replicate the results of the departed graduate student but in fact obtains results contrary to the published work?

Involve students actively in discussions. Students participate most actively when classes are small group discussions rather than large lectures. Discussions allow students to hear other points of view and encourage them to justify their own ideas to others. One technique that is helpful for stimulating discussion is role playing. For instance, in the case about replicating published results, students play the following roles: the new student who cannot replicate the results, the postdoctoral fellow in the laboratory who is familiar with the techniques, the previous student, and the head of the laboratory. If possible, these roles should be assigned before the class session, so that students can think about their positions. To involve other students in the discussion, the teacher can ask them to help one of the students playing a role. The teacher can also stimulate discussion with provocative questions. In Case 1, how practical is it that the scientist learn about the collaborator's discipline? What safeguards can the scientist take against falsification or fabrication of data by the collaborator? What is the community standard regarding responsibility for the part of the work that was done in a collaborator's lab? In the Baltimore/Imanishi-Kari case, a contested issue was the responsibility a scientist takes for the results of experiments carried out in a collaborator's laboratory.

Involve senior scientists in teaching. This is helpful for several reasons. First, it legitimizes concern about ethical issues. The implicit message from senior faculty members whom the students respect as scientists is, "These issues are important to me, and thinking seriously about them is an essential part of being a scientist. Furthermore, if you confront ethical issues in your work, come and talk to me, just as we would talk about any problems with the technical, scientific aspects." At my institution, the University of California San Francisco, the biochemistry department puts on a very successful annual symposium in which senior faculty members take the lead in discussing cases that raised ethical problems for them. Second, senior scientists can present their personal experience with the issues under discussion. One of the most riveting sessions in a symposium our Program in Medical Ethics organized was a panel including a senior faculty member who had been accused of falsifying data and another who had to deal with allegations that one of his fellows had fabricated data. After that session, everyone in the audience learned important lessons: allegations of misconduct need to be taken seriously and that both the accuser and the accused can be harmed by the investigative process.

Offer specific suggestions. In their careers, students are likely to face the ethical issues discussed at this symposium. It is likely that someone in their research group will be accused of misconduct, or that they will be asked to serve on an investigative committee. Discussions of ethical issues should lead to specific suggestions for dealing with the questions that the case poses. In the cases we described, what should the researcher do about looking at the primary electrophysiology data? How should the laboratory director respond when previous work cannot he replicated? The suggestions should be practical, not idealistic responses that overlook the emotions, interpersonal conflicts, institutional constraints, and time pressures that complicate real cases.

In summary, studying ethical issues can result from several motives. Some students attend classes on ethics only because their training grants require them to do so. Other students genuinely believe that these issues are an integral part of scientific training or have supervisors who believe this. Effective teachers need to understand the backgrounds and motives of the students. Starting from the perspective the students bring to the course, we teachers need to try to engage their interest and stimulate them to think about these difficult issues.

Acknowledgement

This paper drew from a project supported by the University of California Systemwide Program in Biotechnology.

The following presentation by Bernard Gert demonstrates one philosopher's approach to raising issues, although it is not the approach for all in the field of philosophy. As can be seen by the subsequent comment and response, the approach to issues in various fields can be strikingly different.

Morality and Scientific Research

Bernard Gert

Any useful attempt to resolve the moral problems that may arise in the course of performing, proposing, or publishing scientific research requires a clear and explicit account of morality. This paper attempts to provide such an account of morality. It is not an attempt to revise our standard or common morality, simply an attempt to describe it. Common morality does not provide a unique solution to every moral problem, but it always provides at least a way of distinguishing between morally acceptable answers and those that are morally unacceptable, i.e., it places a significant limit on legitimate moral disagreement.

Areas of Moral Agreement

One reason for the widely held belief that there is no common or standard morality is that the amount of disagreement in moral judgments is vastly exaggerated. However, everyone agrees that such kinds of actions as killing, causing pain or disability, and depriving of freedom or pleasure are immoral unless one has an adequate justification for doing these kinds of actions. Similarly, everyone agrees that deceiving, breaking a promise, cheating, breaking the law, and neglecting one's duties also need justification in order not to be immoral. No one has any real doubts about this. There is sometimes disagreement about whether a particular act counts as deceiving, but once one has determined that an act counts as deceptive, there is agreement that it needs to be justified. Although there is also some disagreement on what counts as an adequate moral justification for any particular act of deceiving, there is overwhelming agreement on some features of an adequate justification. Everyone agrees that what counts as an adequate justification for one person must be an adequate justification for anyone else when all of the morally relevant features of the situation are the same. This is part of what is meant by saying that morality requires impartiality. There is also agreement that it is not the case that only the most sophisticated philosopher can understand what counts as an adequate justification for deception. No one engages in a moral discussion of questions like "Is it morally acceptable to falsify data in order to increase one's chances of getting a large grant?" because everyone knows that such deception is not justified. Morality is learned by children starting at a very early age, and by the time they are in high school, most children know in most cases whether acting in a certain way is morally acceptable or not. This is part of what is meant by saying that morality is a public system. Finally, everyone agrees that the world would be a better place if everyone acted morally, and that it gets worse as more people act immorally more often. This is why one should try to teach everyone to act morally even though this effort will not be completely successful. Although in particular cases a person might benefit personally from acting immorally, e.g., falsifying data in order to get a grant when there is almost no chance of being found out, even in these cases it is not irrational to act morally, viz., to present the correct data even though it means one will most likely not get the grant. Morality is not determined by feelings or by mystical revelation, it is the kind of public system that every rational person supports. This is part of what is meant by saying that morality is rational.

Rationality as Avoiding Harms and Not Avoiding Benefits

In this section I shall try to provide an account of rationality that explains its relationship to morality and self interest. Everyone agrees that unless one has an adequate reason for doing so, it would be irrational not to avoid any harm or to avoid any benefit. The present account of rationality, although it accurately describes the way in which the concept of rationality is ordinarily used, differs radically from the accounts normally presented in two important ways. First, it starts with irrationality rather than rationality, and second, it defines irrationality by means of a list rather than a formula. The basic definition is as follows: *A person acts irrationally when he acts in a way that he knows (justifiably believes), or should know, will significantly increase the probability that he will suffer any of the items on the following list: death, pain, disability, loss of freedom or loss of pleasure; and he does not have an adequate reason for so acting.*

The close relationship between irrationality and harm is made explicit by this definition, for this list also defines what counts as a harm or an evil. Everything that anyone counts as a harm or an evil, e.g., disease, or punishment, is related to one or more of the items on this list. These items are broad categories, so that nothing is ruled out as a harm or evil which is normally regarded as such. However, except for death, all of these harms have degrees, and even the time of one's death can vary greatly. Further, although there is complete agreement on all of the basic harms, there is no universal agreement on the ranking of these harms. Complete agreement on what the basic harms or evils are, is compatible with considerable disagreement on what is the lesser of two evils.

Since having an adequate reason can make harming oneself rational, a full understanding of rationality requires understanding not only what counts as a reason, but also what makes a reason adequate. The basic definition is as follows: *A reason is a conscious belief that one's action will help anyone, not merely oneself or those one cares about, avoid one of these harms, or gain some good, viz., ability, freedom, or pleasure, and this belief is not seen to be inconsistent with one's other beliefs by almost everyone with similar knowledge and intelligence.* What was said earlier about evils or harms also holds for the goods or benefits mentioned in this definition of a reason. Everything that anyone counts as a benefit or a good, e.g., health, love, or friends, is related to one or more of the items on this list or to the absence of one or more of the items on the list of evils. Complete agreement on what counts as a good is compatible with considerable disagreement on whether one good is better than another, or whether gaining a given good or benefit adequately compensates for suffering a given harm or evil.

A reason is adequate if any significant group of otherwise rational people regard the harm avoided or benefit gained as at least as important as the harm suffered. People are otherwise rational if they do not knowingly suffer any harm without some reason. Just as no beliefs held by any significant religious, national, or cultural group are regarded as delusions or irrational beliefs, e.g., the belief by Jehovah's Witnesses that accepting blood transfusions will have bad consequences for one's afterlife is not regarded as an irrational belief or delusion, so no rankings that are held by any significant religious, national, or cultural group are regarded as irrational, e.g., ranking the harms that would be suffered in an afterlife as worse than dying decades earlier than one would have if one accepted a transfusion is not an irrational ranking. The intent is to not rule out as an adequate reason any relevant belief that has any plausibility; the goal is to provide an account of irrationality on which there is close to universal agreement that no one ever wants anyone he cares for to act irrationally. Only such an account makes irrationality an objective concept and prevents the term "irrational" from degenerating into a term of general disparagement.

This account of rationality, though it may sound obvious, is in conflict with the most common account of rationality, where rationality is limited to an instrumental role. A rational action is often defined as one that maximizes the satisfaction of all of one's desires, but without putting any limit on the content of those desires. This results in an irrational action being defined as any action that is inconsistent with such maximization. But unless desires for any of the harms on the list are ruled out, it turns out that people would not always want those for whom they are concerned to act rationally, e.g., no one wants a loved one who is suffering from a mental disorder to maximize the satisfaction of his desires if this involves self mutilation and suicide.[1] That rationality has a definite content and is not limited to an instrumental role, e.g., acting so as to maximize the satisfaction of all one's desires, goes contrary to most accounts of rational actions.[2]

Scientists may claim that both of these accounts of rationality are misconceived. They may claim that the basic account of rationality does not regard rationality as primarily related to actions at all, but rather regards rationality as reasoning correctly. Scientific rationality consists of using those scientific methods best suited for discovering truth. Although I do not object to this account of rationality, I think that it cannot be taken as the fundamental sense of rationality. The account of rationality as avoiding harms is more basic than that of reasoning correctly, or scientific rationality. Scientific rationality cannot explain why it is rational to avoid suffering avoidable harms when no one benefits in any way. The avoiding harm account of rationality does explain why it is rational to reason correctly and to discover new truth, viz., because doing so helps people to avoid harms and to gain benefits.

It is very important for scientists to realize that the fundamental sense of rationality involves the avoidance of harms, not the seeking of new truth. Not that the seeking of truth is unimportant, we could not avoid harm if we did not know the truth about the world. But to take truth to be the ultimate goal, regardless of the harms involved in obtaining it, is as irrational as the miser taking money to be the ultimate goal. Truth is primarily an instrumental value, although for many of the best scientists it often takes on an aesthetic value as well; they take pleasure in discovering truth for its own sake. However, neither rationally nor morally, does seeking truth confer any special status on an action. Doing scientific research is governed by the same general rational and moral constraints as any other kind of activity.

On the account of rationality I have presented, when there is a conflict between morality and self-interest, it is not irrational to act in either way. Although this means that it is never irrational to act contrary to one's own best interests in order to act morally, it also means that it is never irrational to act in one's own best interest even though this is immoral. Further, it is not irrational to act contrary to both self-interest and morality, e.g., if friends, family, or colleagues, benefit. This latter fact is often not realized, and some physicians and scientists feel that if they act to benefit others and contrary to their own self interest, they cannot be acting immorally. This allows them to immorally cover up the mistakes of their colleagues, believing that they are acting morally, because they, themselves, have nothing to gain and are even putting themselves at risk. Indeed, the misunderstanding of the relationship of loyalty to morality is one of the most important practical issues in moral philosophy.

Although some philosophers have tried to show that it is irrational to act immorally, this conflicts with the ordinary understanding of the matter. Everyone agrees that in some circumstances it may be rational for someone to deceive in order to get a grant, even if this is acting immorally. Nowhere in this paper do I attempt to provide the motivation for a person to act morally. That motivation primarily comes from one's concern for others, together with a realization that it would be arrogant to think that morality does not apply to oneself and one's colleagues in the way that it applies to everyone else. The attempt to provide a useful guide for determining what ways of behaving are morally acceptable when one is confronted with a moral problem presupposes that most of the scientists who read this paper want to act morally.

Morality as a Public System

A public system is a system that has the following characteristics. 1) All persons to whom it applies, those whose behavior is to be guided and judged by that system, understand it, i.e., know what behavior the system prohibits, requires, encourages, and allows. 2) It is not irrational for any of these persons to accept being guided or judged by that system. The clearest example of a public system is a game. A game has an inherent goal and a set of rules that form a system that is understood by all of the players, i.e., they all know what kind of behavior is prohibited, required, encouraged, and allowed by the game; and it is not irrational for all players to use the goal and the rules of the game to guide their own behavior and to judge the behavior of other players by them. Although a game is a public system, it applies only to those playing the game. Morality is a public system that applies to all moral agents; all people are subject to morality simply by virtue of being rational persons who are responsible for their actions.

In order for morality to be known by all rational persons, it cannot be based on any factual beliefs that are not shared by all rational persons. Those beliefs that are held by all rational persons include general factual beliefs such as: people are mortal, can suffer pain, can be disabled, and can be deprived of freedom or pleasure; also people have limited knowledge, i.e., people know some things about the world, but no one knows everything. On the other hand, not all rational people share the same scientific and religious beliefs, so that no scientific or religious beliefs can form part of the basis of morality itself, although, of course, such beliefs are often relevant to making particular moral judgments. In a parallel fashion, only personal beliefs that all rational persons have about themselves, e.g., beliefs that one can be killed and suffer pain, etc. can be used in providing a foundation for morality. Excluded as part of a foundation for morality are all personal beliefs about one's race, sex, religion, etc., because these beliefs are not shared by all rational persons.

Although morality itself can be based only on those factual beliefs which are shared by all rational persons, individual moral decisions and judgments obviously depend not only on the moral system, but also on the situation. In fact, most actual moral disagreements, e.g., whether a particular scientist acted properly or not, are based on a disagreement on the facts of the case, e.g., whether or not he knowingly used information that he gained from reviewing a grant to formulate his own grant proposal. Other moral disagreements depend upon disagreements on what standards are appropriate to apply, e.g., should a competent scientist have known that the experiment posed a significant risk of harm? Still others may depend upon disagreements about what counts as breaking a rule, e.g., when does not reporting failed experiments count as deception. Thus in order to be qualified to make a moral judgment in a particular field, one must know the conventions and practices of that field.

Almost all the difficult moral cases that arise in the area of scientific research depend upon determining whether the action or practice under consideration is one that needs justification, e.g., is deceptive. Unlike the field of medicine, where determining that the action needs justification is often just the start of the moral inquiry, e.g., one has to decide whether the benefit to the patient justifies deceiving him, in scientific research, it is only rarely that people attempt to justify what they acknowledge to be deceptive, e.g., deception experiments in psychology. Generally, the issue turns completely on whether or not the behavior is properly characterized as deceptive, e.g., using a non-standard statistical method that produces more significant results or listing as an author someone who has not even seen the paper. Thus for most moral issues that arise in scientific research, there is no need for a sophisticated philosophical account of morality. In science, everyone acknowledges that deception is unjustified and the whole discussion turns on whether or not the action or practice counts as deceptive. That is why knowledge of the field is so important, for most of the reasoning involved concerns whether or not people are misled or deceived by the action or practice. Here, knowledge of what scientists and consumers of science believe is more useful than knowledge of morality. The framework of our common morality can be applied to a scientific practice only after it is clear how a particular act or practice should be characterized. However, knowledge of the moral framework is helpful in showing that there is no special morality for scientists.

Although morality is a system that is known by all those who are held responsible for their actions, it is not a simple system. A useful analogy is the grammatical system used by all competent speakers of a language. Almost no competent speaker can explicitly describe this system, yet they all know it in the sense that they use it when speaking themselves and in interpreting the speech of others. If presented with an explicit account of the grammatical system, competent speakers have the final word on its accuracy. They should not accept any description of the grammatical system if it rules out speaking in a way that they regard as acceptable or allows speaking in way that they regard as completely unacceptable.

In a similar fashion, a description of morality or the moral system that conflicts with one's own considered moral judgments normally should not be accepted. However, an explicit account of the systematic character of morality may make apparent some inconsistencies in one's moral judgments. Moral problems cannot be adequately discussed as if they were isolated problems whose solution did not have implications for all other moral problems. Providing an explicit account of morality may reveal that some of one's moral judgments in one area are inconsistent with the vast majority of one's other judgments. Thus one may come to see that what was accepted by oneself as a correct moral judgment is in fact mistaken. Even without challenging the main body of accepted moral judgments, particular moral judgments, even of competent people, may sometimes be shown to be mistaken, especially when long accepted ways of thinking are being challenged. In these situations, one may come to see that one was misled by superficial similarities and differences and so was led into making judgments that are inconsistent with the vast majority of one's other moral judgments. For example, many scientists have recently discovered that their moral judgments about what was morally allowable regarding who should be listed as an author of an article are inconsistent with the vast majority of their other moral judgments.

As noted earlier, there are certain kinds of actions that everyone regards as being immoral unless one has an adequate justification for doing them. These kinds of actions are killing, causing pain or disability, depriving of freedom or pleasure, deceiving, breaking a promise, cheating, breaking the law and neglecting one's duty. Anyone who kills people, causes them pain, deceives them, or breaks a promise, and does so without an adequate justification, is universally regarded as acting immorally. Saying that there is a moral rule prohibiting a kind of act is simply another way of saying that a certain kind of act is immoral unless it is justified. Saying that breaking a moral rule is justified in a particular situation, e.g., breaking a promise in order to save a life, is another way of saying that a kind of act that would be immoral if not justified, is justified in this kind of situation. When no moral rule is being violated, saying that someone is following a moral ideal, e.g., relieving pain, is another way of saying that he is doing a kind of action regarded as morally good. Using this terminology allows one to formulate a precise account of morality, showing how its various component parts are related.[3] Such an account may be helpful to those who must confront the problems raised by the practices that they encounter in performing, proposing, or publishing scientific research.

Justifying Violations of the Moral Rules

Almost everyone agrees that the moral rules are not absolute, that all of them have justified exceptions; most agree that even killing is justified in self-defense. Further, one finds almost complete agreement on several features that all justified exceptions have. The first of these involves impartiality. Everyone agrees that all justified violations of the rules are such that if they are justified for any person, they are justified for every person when all of the morally relevant features are the same. The major, and probably only, value of simple slogans like the Golden Rule, "Do unto others as you would have them do unto you" and Kant's Categorical Imperative, "Act only on that maxim that you would will to be a universal law" are as heuristic devices designed to get one to act impartially when one is contemplating violating a moral rule. In trying to decide what to do in difficult cases, however, it is more useful and less likely to be misleading to consider whether one would be prepared to publicly allow that kind of violation, i.e., allow it to be included in that public system which is morality.

Acting in an impartial manner with regard to the moral rules is analogous to a referee impartially officiating a basketball game, except that the referee is not part of the group toward which he is supposed to be impartial. The referee judges all participants impartially if he makes the same decision regardless of which player or team is benefited or harmed by that decision. All impartial referees need not prefer the same style of basketball; one referee might prefer a game with less bodily contact, hence calling more fouls, while another may prefer a more physical game, hence calling fewer fouls. Impartiality allows these differences as long as he does not tell only one team of these preferences and does not favor any particular team or player over any other in calling fouls. In the same way, moral impartiality allows for differences in the ranking of various harms and benefits as long as one would be willing to make these rankings public and one does not favor any one rational person or group of persons, including oneself and one's friends, over any others when one decides whether to violate a moral rule or judges whether a violation is justified.

The next feature on which there is almost complete agreement is that it has to be rational to favor everyone being allowed to violate the rule in these circumstances. Suppose that some person suffering from a mental disorder both wants to inflict pain on others and wants pain inflicted on himself. He favors allowing all persons to cause pain to others if they would not complain if others caused pain to them. This is not sufficient to justify that kind of violation. No impartial rational person would favor allowing those who don't complain when they are caused pain to cause pain to everyone else. The result of allowing that kind of violation would be an increase in the amount of pain suffered with no benefit to anyone, which is clearly irrational.

Finally, there is general agreement that the violation be publicly allowed, i.e., that one favors the violation even if everyone knows that this kind of violation is allowed. It is not sufficient to justify a violation that it would be rational to favor allowing everyone to violate the rule in the same circumstances, if one favors it only if almost no one knows that it is allowable to violate the rule in those circumstances. For example, when almost no one knows that such deception is allowed, it might be rational for one to favor allowing a great scientist to deceive others by claiming to have greater confirmation than he actually has if failing to make this false claim is likely to lead other scientists to give up on a theory that he is absolutely convinced is true. But that would not make deception in these circumstances justified. It has to be rational to favor allowing this kind of deception when everyone knows that it is allowed to deceive in these circumstances. The requirement that the violation be publicly allowed guarantees the kind of impartiality required by morality.

Not everyone agrees on which violations satisfy the three conditions of impartiality, rationality, and publicity, but it is part of our moral system that no violation is justified unless it satisfies all three of these conditions. Acknowledging the significant agreement concerning justified violations of the moral rules, while allowing for some disagreement, results in the following formulation of the appropriate moral attitude toward violations of the moral rules: *Everyone is always to obey the rule unless an impartial rational person can advocate that violating it be publicly allowed. Anyone who violates the rule when no impartial rational person can advocate that such a violation be publicly allowed may be punished.* (The 'unless clause' only means that when an impartial rational person can advocate that such a violation be publicly allowed, impartial rational persons may disagree on whether or not one should obey the rule. It does not mean that they agree one should not obey the rule.)

Anyone acting or judging as an impartial rational person decides whether or not to advocate that a violation be publicly allowed by estimating what effect this kind of violation, if publicly allowed, would have. However, rational persons, even if equally informed, may disagree in their estimate of whether more or less harm will result from this kind of violation being publicly allowed. Disagreements in the estimates of whether a given kind of violation being publicly allowed will result in more or less harm may stem from two distinct sources. The first is a difference in the rankings of the various kinds of harms. If someone ranks a specified amount of pain and suffering as worse than a specified amount of loss of freedom, and someone else ranks them in the opposite way, then although they agree that a given action is the same kind of violation, they may disagree on whether or not to advocate that this kind of violation be publicly allowed. The second is a difference in estimates of how much harm would result from publicly allowing a given kind of violation, even when there seems to be no difference in the rankings of the different kinds of harms. These differences may stem from differences in beliefs about human nature or about the nature of human societies. In so far as these differences cannot be settled by any universally agreed upon empirical method, such differences are best regarded as ideological. But sometimes there seems to be an unresolvable difference when a careful examination of the issue shows that there is actually a correct answer.

Applying Morality to a Particular Case

Some scientists may claim that certain kinds of deception, e.g., some kinds of misreporting of their experiments, in order to enhance the acceptability of an hypothesis of whose correctness one is very confident, is justified. They may hold that deception in these circumstances conflicts less with the successful practice of science than if such enhancement were not practiced. They may hold that if experienced scientists are very confident of their claims, those claims are usually true, so that such enhancement will result in less time being wasted doing futile research. Thus they may claim that this kind of deception actually results in more truth being discovered than failure to deceive. It may be, in the words of the COSEPUP Panel on Scientific Responsibility and the Conduct of Research that published *Responsible Science: Ensuring the Integrity of the Research Process*, (National Academy Press, 1992) a questionable practice, but it is not scientific misconduct.

In what follows I shall apply the account of morality to a particular example of a questionable practice. I do not claim that as presently understood, what was done would count as scientific misconduct, that is, it was not fabrication, falsification, or plagiarism. But I hold that, from the point of view of morality, the only difference between many questionable practices and scientific misconduct is that the former are fairly widespread, the later fairly rare. Further, I find it far more interesting and important to discuss questionable practices, for if I can persuade a sufficient number of scientists that a questionable deceptive practice ought to be discontinued, I will have done the very small amount a moral philosopher can do toward the advancement of science.

I am primarily interested in the questionable practice of misleadingly enhancing the strength of one's experiments, no matter how this is done, e.g., using non-standard statistical techniques, or withholding disconfirming data. The case I shall discuss is nominally that of Robert Andrews Millikan and the oil drop experiments. But I am not interested in the historical Millikan, only with that Millikan presented in an article entitled "Scientific Fraud" by David Goodstein. (*The American Scholar*, Volume 60, Number 4, Autumn 1991). All of the information I have on this case comes from reading that article. Professor Goodstein wrote this article in response to Broad and Wade, and it was written with an irony that was probably quite obvious to his scientific readers. Professor Goodstein did not intend all of the statements in his article to be taken literally, but the article expresses very clearly views that are actually held by some scientists, though often much less articulately expressed. Those scientists who hold these views usually do so because they implicitly hold a philosophical view that is known as consequentialism, the view that morally speaking, all that counts are the consequences of one's actions. Thus I shall call the person who literally holds all of the views expressed in the paper, "the consequentialist," and I shall argue as if the consequentialist wrote the paper.

The consequentialist claims that "if all scientists rigorously adhered to proper scientific procedure at all times, very little scientific progress would occur." (p. 509) Thus, the consequentialist seems to be and, as we shall see later, actually is advocating that scientists, especially great scientists, not always adhere to proper scientific procedure.

The example that the consequentialist discusses most fully is that of Robert Andrews Millikan and the oil drop experiments, although he also defends Newton. Some scientists seem to react very much like some sports fans, if a scientist or player is great enough, then almost nothing they do counts as seriously wrong, if it is wrong at all, just as some sports fans think that a truly great player is entitled to act in normally unacceptable ways. I heard a respected scientist talk about behavior that he and everyone else would regard as unacceptable behavior in the lab, e.g., disrupting someone else's experiment for one's own convenience, in a way that was interpreted by several people who heard him, including me, as acceptable when the scientist behaving in that way is of sufficient stature.

The excessive hero worship of great scientists, the often unconscious assumption that a good scientist must also be a morally good person, together with the standard loyalty of members of a profession to one another, may explain, although it does not justify, the failure to acknowledge that great scientists have sometimes behaved immorally. Indeed, they have sometimes behaved immorally when doing science. That a great scientist is not a saint, that he has succumbed to temptation on occasion, does not mean that he is a bad person. All of us have behaved immorally from time to time, it is worse than pointless to maintain that great scientists have never done so. The consequentialist says, "The Myth of the Noble Scientist serves us poorly precisely because it obscures the distinction between harmless minor hypocrisies and real fraud."(p. 515) But unfortunately, the way that the consequentialist makes this distinction is that the former are done by really great scientists who turn out to have been on the right track, and the latter are done by more ordinary scientists who were probably wrong anyway. The consequentialist comes very close to recognizing that Millikan acted improperly and yet his view prevents him from completely admitting it. The following paragraph is worth quoting in its entirety.

It is worth noticing in these instances that both Newton and Millikan were motivated by the need to convince a skeptical world of what they perceived to be scientific truth. In both cases inestimable damage to science would have been done had they not succeeded. Nevertheless, as noted earlier, perpetrators of real fraud also generally do so when they are convinced that they know the right answer to the scientific question they are investigating. Newton and Millikan did not commit fraud, and what they did was necessary and important, but they shared something distressingly in common with those who have been truly guilty. (p. 512)

The consequentialist does not seem to realize that what they shared "distressingly in common," is that they deceived others about their work. They withheld information that they knew would undermine their purpose of convincing "a skeptical world of what they perceived to be scientific truth," (p. 512) and in Millikan's case the article makes it seem as if he simply lied. Their legacy and defense provides the climate in which other scientists justify to themselves their acts of deception. How does the consequentialist know that "inestimable damage to science would have been done had they [Newton and Millikan] not succeeded?" How does he know that if Newton and Millikan had been more straightforward, admitting that their results were not perfect, that others might not have acted in such a way as to bring about the correct results more quickly? How does he know that these acts of honesty under great temptation to deceive might not have so influenced the scientific tradition that science would be even more successful than it now is? He does not and cannot know these things, and the claim about inestimable damage to science is simply made in order to distinguish between the behavior of these great scientists and the behavior of lesser scientists who acted improperly when there is no morally relevant distinction between the two cases at all.

Let us examine the case of Millikan in more detail. I do not challenge the consequentialist's account of the facts at all, for not only is this the only account of the facts that I have, but even more important, I base my judgment that Millikan acted improperly on the consequentialist's own account of the facts. According to the consequentialist, "Millikan was measuring the electric charges of oil drops; he wanted to prove that the electron charge came in definite units... Millikan had a rival, Felix Ehrenhaft, who believed that the electric charge was a continuous rather than a quantized quantity. Ehrenhaft criticized Millikan's results, so Millikan went back to the laboratory to get better data to have ammunition against Ehrenhaft. Later on Millikan published a paper in *Physical Review* in which he says he is publishing data from 'all of the drops experimented upon during 60 consecutive days.' " (p. 511)

Millikan and Ehrenhaft are having a dispute; Millikan does experiments to show that he is right and publishes an article saying that all of the data support his view. But as the consequentialist notes, "Millikan's notebooks, however, tell a different story." (p. 511) The data were not quite so consistent as Millikan claimed. "He knew, of course, what result he expected. So in some cases he would write in red (everything else is black), 'Beauty-Publish,' or 'One of the best I've ever had — Publish.' And then on one page he wrote, 'Very low — something wrong.' And, of course, that one did not get published, in spite of the fact that he said he published everything." (p. 511) There is no indication that prior to getting his unexpected result, Millikan knew that the experiment had gone wrong. Quite the contrary, it was because he got a result he didn't want that he concluded that something had gone wrong. Publishing only the data that support your conclusion is bad enough, explicitly claiming that you are publishing all the data when you are not is out and out deception.

Imagine that we are being told exactly this kind of story of a scientist's behavior, except that it is a graduate student in one's lab, not Millikan, that has published the false statement. What would be the consequentialist's reaction to this student's behavior? As a consequentialist should he wait to see whether the results were right or wrong before he decided to congratulate or condemn the behavior? I would hope that most scientists would make clear to the student that this kind of behavior was not acceptable, that if it happened again he would be dropped from the program. I would hope that they would explain to the student why such behavior is unacceptable, even when one is very confident of one's results. But if a scientist accepts what the consequentialist has written, I am not confident that my hopes would be realized. Unfortunately, my experience indicates that the consequentialist's views are shared by many scientists.

Editors of scientific journals are not even prepared to include as a footnote to articles, some phrase like the following: "Some experiments gave results which deviate significantly from those published, but we believe that

these experiments were defective. Write us if you want further details."
This phrase would not be used about experiments which you knew were defective and decided not to use prior to getting the results, but only for those
experiments in which the results are not used only after it is discovered that
they are out of line with other experiments performed. Publishing such a footnote, which Millikan could have done, would not unduly clog up the journals.
The consequentialist's remark, "If every scientist were obliged to publish
every mistake, the literature would be so full of garbage, it would be unread-
able (it's bad enough as it is)." (p. 511) is a standard rhetorical ploy that is
completely irrelevant.

It is not the clogging of the journals that is the consequentialist's primary
concern. "Even worse, in Millikan's case, any mistake would seem like confirmation of Ehrenhaft's contention." (p. 511) Since Millikan was right, his
behavior was acceptable; the end, scientific progress justifies immoral means,
deception. The consequentialist continues, "So when he got a wrong result,
or when he could observe directly that a drop was not behaving properly, he
would examine his apparatus to find his mistake so that he could fix it. It
didn't count as one of the ' drops experimented upon.' Needless to say, when
he got a result that agreed with his expectation within his expected limits of
error, he did not try very hard to find some reason for throwing it away. This,
too is accepted behavior, even though it builds a real bias unto the results.
Millikan was not committing fraud. He was exercising scientific judgment."
(pp. 511-512)

Notice how incredible this view is. Imagine teaching a graduate student
to exercise his scientific judgment by keeping secret all unwanted results,
publishing only those that confirmed his hypothesis, and claiming that he is
publishing everything. Imagine holding that it is accepted scientific procedure
to behave in a way that one knows "builds a real bias into the results." If this is
the way in which graduate students in science are being taught, it is amazing
that scientific fraud is not far more common than it seems to be. But I suspect
that the students are made to understand, though not explicitly told, that this
kind of behavior is not acceptable for them, it is only acceptable for someone
who has already attained stature in the field.

The consequentialist's double standard, one for great scientists, the other
for the rest of us, obviously troubles him. He does not want to hold a double
standard because he knows that students learn by imitation, so that it would
have bad consequences to tell students to behave in a certain way and then
have them watch the great scientist who does not behave in that way. Thus, he
tells us what would have been unacceptable behavior for Millikan. "Millikan
did not simply throw away drops he didn't like. That would have been fraud by
any scientist's standard. To discard a drop he had to find some mistake that
would invalidate that datum ('distance wrong,' he wrote on a page)." (p. 511)
It is commendable that Millikan felt obliged to find a mistake whenever the
results turned out wrong, but that he did not commit what the consequentialist
calls "fraud by any scientist's standard" does not mean that he did nothing

wrong. The consequentialist admits that "Millikan's notebooks tell a different story" than Millikan's claim that he is publishing data from 'all of the drops experimented upon during 60 consecutive days'." (p. 511)

The consequentialist, like most other scientists, is genuinely concerned with the progress of science. Most members of every profession are genuinely concerned with the goals of their profession. Most members of a profession, like most scientists, think that those who are not members of their profession do not really understand their profession. They are also very reluctant to have anyone else set rules for their profession, even when they know that their profession is doing some things poorly. Thus the consequentialist says, "Some people object to guest authorship (putting the boss's name on the paper even though the boss did not participate in the research), but the practice appears to be perfectly standard in some fields." (p. 513) But, "perfectly standard" can only mean commonly done, it does not mean that it is acceptable behavior. Why would the boss want his name on the paper unless it indicated that he had done some work on it? Otherwise, a note as to whose laboratory the work was done in would be completely satisfactory. The "perfectly standard" practice of guest authorship can only be a form of deception. But since the boss probably has already gained stature in the field, this may make his behavior acceptable. The consequentialist's concern with protecting members of the profession, especially the great scientists, comes out clearly in the last sentence of the article. "I can only hope that we won't wind up arranging things in such a way as would have inhibited Newton or Millikan from doing their thing." (p. 515) I do not want to inhibit scientists from exercising their scientific judgment, all I want is that they not deceive others about what they are doing.[4]

Some scientists do claim that the kind of deception practiced by Millikan, no matter how experienced the scientist or how confident he is of his claim, is not justified. However, they usually also base their view on what benefits scientific progress. They hold that deception of this kind will actually increase the amount of futile research, because, e.g., other scientists may come to know of such practices and so have less faith in the claims of other scientists, thus not accepting their claims even when they have been genuinely confirmed. Thus there is a genuine empirical dispute about whether or not deception in the reporting of disconfirming evidence by experienced scientists in order to enhance the acceptability of hypotheses of whose correctness they are very confident will increase or decrease the amount of futile research. I do not know which of these empirical claims about the actual effect of deception is correct, but if we are concerned with the moral justifiability of deception it does not matter.

The morally decisive question is "What would be the consequences if this kind of deception were publicly allowed?" Consequentialists do not take into account that a justifiable violation of the rule prohibiting deception must be one that is publicly allowed, which means that everyone must know that this kind of deception is allowed. Once one realizes that if everyone knows that it is allowable to deceive, e.g., to enhance the acceptability of one's confidently held claims, then the loss of trust involved will obviously have a worse result than if everyone knew that such deception was not allowed. It is only by limiting one's attention to the results of one's own deception that one could be led to think that such deception was justified. Acknowledging that a violation is morally allowed for oneself only if one is willing that everyone know that this kind of violation is allowed for everyone, makes clear the deception is morally unjustified. Consciously holding that it is morally acceptable for oneself to deceive in this way, although, of course, it would not be acceptable for everyone to be allowed to deceive in the same circumstances, is exactly what is meant by arrogance, the arrogating of exceptions to the moral rules for oneself that one would not allow for others. This arrogance is clearly incompatible with the kind of impartiality that morality requires with regard to obeying the moral rules.[5]

References

1. See "Irrationality and the DSM-III-R Definition of Mental Disorder," *Analyse & Kritik*, Jahrgang 12, Heft 1, Juli 1990, pp. 34-46

2. See "Rationality, Human Nature, and Lists," *Ethics*, Vol. 100, No. 2, January 1990, pp. 279-300

3. See *MORALITY: A New Justification of the Moral Rules*, Oxford University Press, 1988, 317 pp., paperback, 1989, for a fuller account.

4. The relevance of this discussion to contemporary problems is clear from the report on University of Pittsburgh lead researcher, Herbert Needleman. See *The Journal of NIH Research* (December 1992, Vol. 4 No.12, p. 44). According to reports, "Needleman deliberately misrepresented his procedures in the study published in the March 29, 1979 issue of the *New England Journal of Medicine*. But it [The University of Pittsburgh hearing board] nevertheless finds him not guilty of scientific misconduct." As with Millikan, it turns out that Needleman's conclusions were correct. "Indeed, the report says that three reanalyses of Needleman's data, including one by members of the board, all agree that the evidence in support of the Needleman's hypothesis is stronger than reported in the 1979 paper." However, "The 'board was unanimous in its belief that Needleman was deliberately misleading in the published account' of the selection criteria he and his colleagues used, the report says." "For motive, the board suggests that 'the misrepresentations may have been done to make. . . the procedures appear more rigorous then they were, perhaps to ensure publication.' " "The board's decision, in effect, was that lying in a published report of research is not scientific misconduct." It seems clear that Needleman, like Millikan, lied in order to gain support for an hypothesis that he believed in and in fact turned out to be correct. The Office of Research Integrity is now reviewing the case, so that the final decision has not yet been made.

5. This work has been supported in part by funds from NIH (HG00130). They of course have no responsibility for its content.

A **Comment** by David L. Goodstein

Professor Gert believes that all scientists are hero-worshiping consequentialists. That is to say, he believes that scientists judge the morality of their behavior only on the basis of whether the outcome was correct or not, and that we have different standards of behavior for great scientists and for lesser ones. Anyone intrepid enough to have read all the way through Professor Gert's article has learned first of all that the word "theory" has a radically different meaning in the phrase "moral theory" from what it does in the phrase "scientific theory," and secondly that he has chosen to reveal his views about scientists by applying his moral theory to a few paragraphs about Millikan in a semi-popular article about scientific fraud that I wrote a few years ago.

Gert begins his analysis by telling us that he has no interest in the real, historical Millikan. In other words, facts will not be permitted to interfere with his theory. He goes on to admit that everything he knows about the subject he learned from those few paragraphs in my article. Furthermore, he admits up front that he is purposely going to misrepresent even that little bit ("Professor Goodstein did not intend all of the statements in his article to be taken literally..."). He seems to think that by telling us at the outset that he is going to deceive the reader by misrepresenting my views, he absolves himself of all guilt for having done so. Thus he is able to use words quoted from my article to represent the prototypical scientist that he calls "the consequentialist," desperately in need of Gert's moral guidance.

This bizarre behavior on Gert's part came about because detecting irony and humor in written English is not one of his intellectual strengths. When he was made to understand that in an earlier and even more remarkable draft of his paper he had hilariously misunderstood my poor three paragraphs, he was unable to give up his incisive analysis, so he decided to proceed by telling the reader that he was going to pretend I meant to say what he so badly wanted to believe I had said.

These words "misrepresent" and "deceive," which accurately apply to Gert's essay would be the key elements necessary to prove misconduct in science. I do not mean to accuse Gert of misconduct; that word should be reserved for more important matters. I do think, however, that his article proves conclusively that we scientists have nothing to learn from him about ethics or morality. I'm afraid we're just going to have to muddle through on our own.

A Response by Bernard Gert

I am very sorry to have upset Professor Goodstein. The convention in moral philosophy is to be concerned with the arguments that are put forward to defend certain kinds of behavior, whether that behavior actually occurred or is merely hypothetical. We are also unconcerned with being mentioned in connection with these arguments, especially if it is not being claimed that we have endorsed them. The conventions in physics are clearly quite different, and I apologize to Professor Goodstein for not becoming more informed about the relevant conventions in physics.

I also hold, although I am not sure that it is a general convention in philosophy, that one ought to try to be as careful when writing for the general public as when writing for a scholarly journal. For understandable reasons that may not be a convention in physics.

I am disturbed that I am accused of violating the moral rule against deceiving that I so strongly endorse in my paper. In that paper I explicitly state, "I do not want to inhibit any scientist from exercising his scientific judgment, all I want is that he not deceive anyone about what he is doing." Professor Goodstein's comments make it quite clear that I was not trying to deceive anyone about what I was doing.

Ethics and Values in Science: The NAS-COSEPUP Report on Scientific Responsibility and the Conduct of Research

Edward E. David, Jr.

I. Introduction

Malfeasance associated with science is not a new phenomenon. There have been episodes and cases that reach back to the 19th century and beyond. The Piltdown man fabrication was discovered only a decade or so ago. Apparently it was a hoax perpetuated by an unknown person. Piltdown man was a recognized part of the fossil record for many years before the fake was discovered. The existence of N-rays was put forward by a French investigator at the time when alpha, beta, and gamma radioactive rays were being established. It turned out that N-rays didn't exist, though alpha, beta, and gamma rays did. The debunking of N-rays is a story in itself. Cyril Burt, though deceased, has been accused of fabricating data in his paper about the comparative mental abilities of racial populations. Even data supporting Millikan's oil drop experiment have been called into question.

The Summerlin case in 1974 seems to have been the beginning of current concerns. A physician-researcher at Sloan-Kettering Institute apparently painted spots with a black marker pen on the skins of mice to prove efficacy of a skin graft technique. There have been a series of other episodes, culminating in what has been called the Baltimore case. Congressman Dingell and Dr. David Baltimore contended over this matter with wide publicity. Another well-known affair involves Dr. Robert Gallo of NIH, and credit for discovery of the AIDS virus. There have been recent developments in this case as reported in *Science* magazine of January 8, 1993.

The effect of these happenings has been to raise the misconduct-in-science issue to a major subject for the National Academy of Sciences, not to mention the scientific community as a whole. The NAS President, Frank Press, has said in part, "This issue can't be swept under the rug." Indeed, it is a serious matter, not only when Federal funds are involved but also whenever the integrity of the research process is called into question.

Science today is a major resource in policy-making, legislation, and regulation at every level of government and in operating and planning at entrepreneurial corporations and even stodgy ones. If the credibility and integrity of research is undermined, the public, the economy, and governance itself could suffer. Protecting the integrity of research is the proper goal of any steps to avoid misconduct in science.

II. The NAS-COSEPUP Panel

This situation was well noted by the Academy of Sciences' Committee on Science and Engineering Public Policy (COSEPUP). After lengthy and considered discussions, COSEPUP agreed to sponsor a full-scale review. It established a panel to study current research practices, to assess the system of self-governance in research, and to recommend proper roles for public and private institutions in assuring research integrity. The panel was asked also to examine the pros and cons of formal guidelines for responsible research practices.

This assignment raises a number of fundamental questions, some of them quite vexing.

- How widespread is misconduct in science?

- What are the contributing causes?

- How should cases be handled when they arise? What procedures could provide protection of individuals, especially whistle blowers, and provide due process for all?

- What role should education of students and research people play? Are students the right audience or do senior people require education, too? Note that most confirmed instances of misconduct in science involve senior people, not students or juniors.

The panel membership was intentionally diverse. It included a broad range of members. Among them were bench scientists, science policy people, university and industrial administrators, professors, whistle blowers, philosophers, lawyers, and politically-oriented people. There were industrial leaders, academic leaders, and the young and old. There were 22 panel members in all. While the panel could not be called a formal representative of the science community, it was broad enough to be highly credible. Its recommendations should have standing and should carry weight with those who would have to carry them out. The panel members brought a wide range of conflicting viewpoints, but they managed to achieve consensus on important issues, although two members refused to sign the final report.

III. Summary of Results

The problem is real and must be addressed. The problem was not conjured up by political forces. Steps are needed to preserve self-governance by the scientific community, to assure and reinforce the integrity of the research process, and to avoid measures such as research policing by enforcement or auditing authorities.

Steps to address the problem were recommended by the panel, recognizing that there is *no magic bullet*. Rather a number of steps by a variety of government-academic-laboratory institutions are required. Some of these steps are:

1. Education of researchers and practitioners on what is expected,

2. Responsibility for integrity of research to be assumed by institutions and laboratories,

3. Establishment of a non-government, privately-funded board to collect and disseminate information on the state of research ethics,

4. The Federal government to adopt a single definition of misconduct in science so that practitioners know unambiguously what behaviors are expected and those that are excluded.

IV. What is Misconduct in Science?

In adopting a working definition of misconduct in science, the panel avoided a "know-it-when-you-see-it" definition, as has been used in judging what is pornographic. The panel also avoided "defining the problem away." One panelist said that misconduct in science is an oxymoron since misconduct cannot be a part of science. If it is misconduct, it is not science. True enough, but what about perversions of science that occur? What are these?

There are three "cardinal sins" that were recognized as the crux of misconduct in science:

• Fabrication of data or results

• Falsification in reporting on research not done or changing data or results

• Plagiarism - appropriating *without proper credit* not only others' words, but also their ideas.

These three sins are abbreviated and called *FFP*: So, misconduct in science is fabrication, falsification, or plagiarism in proposing, performing, or reporting research. Note that intent by perpetrator is not implied. Proving intent is a diversion from finding facts and acting accordingly.

Current formal definitions of misconduct in science by NSF and NIH include FFP, but also an addition; namely, "other serious deviations from accepted research practices." The ambiguity of this phrase is likely to cause more controversy than enlightenment since accepted practices are different in various disciplines and laboratories. Furthermore, creative science often departs from accepted practice. Note that almost all proven cases to date of misconduct in science involved FFP. The favored definition does not include errors of judgment, errors in recording, selection, or analysis of data, differences in opinions, or misconduct unrelated directly to the research process.

The latter recognizes that *misconduct in science is not the same as misconduct by scientists*. Furthermore, even though a scientist may take actions which are not admired, such as not maintaining adequate research records, inadequately supervising subordinates or exploiting or harassing them, or senior people taking credit for juniors' work, these do not fall under FFP. Rather they are classed as *questionable research practices*. These should be addressed through the promotion and reward systems of institutions and should be further discouraged by educational programs. This category (questionable research practices) gives the flexibility to account for behaviors in the context of what is being accomplished. Traditionally, institutions are willing to tolerate more contrary behavior in productive people than in those who are only objectionable. But, questionable research practices should be discouraged because they are a poor model for students and subordinates, and because they can verge into FFP.

There can be other misbehaviors by scientists; misappropriation of funds, rape, arson, trespass, and so on. If we group these as "other misconduct," then we have a logically complete set of definitions; namely, misconduct in science (FFP), questionable research practices by scientists, and other misconduct. Panelist, Howard Schachman has pointed out that any definition should separate the "crooks" from the "jerks." This system will serve that need.

Institutions and governments should concern themselves with FFP and handle these episodes by mechanisms crafted for the purpose. *Questionable research practices* should be discouraged by educational programs as well as by the usual social disapproval and career penalties, except in flagrant cases such as sexual harassment, which can be prosecuted under existing laws or regulations. Cases of *other misconduct* should be addressed by policies that apply to all institutional members, not just scientists. In setting up mechanisms to find facts and investigate accusations of FFP, institutions must realize that it is a daunting task to question the veracity and actions of prominent people. One remedy is to have procedures in place prior to conducting an investigation. Another is to use investigators outside the institution, as is done for academic visiting committee appointments.

V. The State of Ethics in Science and Causes of Misconduct

The Panel found that there is little data on misconduct in science cases, except in a few prominent instances. Conclusions about the extent of misconduct in science are mostly speculation and surmise. It is clear that misconduct in science arises not just from bad apples. Institutional factors probably contribute as well.

The NSF and NIH have offices that respond to accusations pointed at their grantees, contractors, and intramural staff. NSF and NIH report that from 1989 to 1991, some 200 allegations of misconduct in science were reported. More than 30 cases have so far been confirmed as valid. Some cases are still under investigation. We don't know if the frequency of such incidents is increasing or decreasing. We don't know about the frequency or seriousness of incidents in industry or in other countries.

Nevertheless, any level of misconduct is damaging to science and to the public, and there is the feeling that the incidence is increasing. If so, what are the contributing causes? Here, we get heavily into speculation. Among the causes put forward are:

1. People not trained in research methods, especially in the medical sciences, are major causes. It is true that many of the known cases have arisen in the medical sciences. But, there is no evidence that they are peculiarly susceptible to misconduct in science. Cases arise *not* rarely in the physical and social sciences.

2. The diversity of people in research laboratories is increasing. These diverse cohorts come from different cultures with different ethics and morals, so the lab atmosphere becomes contentious and each group's expectations are violated. Again, there is no evidence to support this assertion.

3. The increasing competitive pressures to succeed in preparing winning proposals and in attracting research funding in competition with other scientists are thought to encourage misconduct in science. Again, there is no supporting evidence beyond anecdotal stories. Nevertheless, there is the temptation to use this thinking to say that more research money would solve the misconduct problem. The panel rejected this idea.

In my view, we just don't know what is behind misconduct instances. Probably the causes are complex, and often specific to the episode. A too-neat separation of causal influences, yet one that is often voiced is (1) the "bad apple" theory; (2) the proposal-grant-contract system breeds misconduct. The panel didn't believe either was a comprehensive cause, but did emphasize that although the system is not rotten, some reforms are necessary.

VI. What Is To Be Done?

Note that three traditional mechanisms are protection against FFP. Two of these are peer review of research results and publication review of manuscripts. These continue to be effective in some instances, but by themselves are no longer completely adequate, according to our panel. The third traditional protection is the so-called self-correcting nature of science, which hinges on repetition of research results. This protection, too, is no longer adequate, if it ever was. Without going into detail, just note that planetary data gathering, megamouse experiments, and extensive epidemiological studies are not easily susceptible to repetition. Long intervals may pass before corrective repetitive experiments can be done, meanwhile fallacious results can cause severe damage.

There are several ways of augmenting peer and publication reviews. Among them are assigning institutional responsibility for the integrity of the research process, which in turn may include institutional or laboratory guidelines for research practices, educational programs to clarify ethical expectations, and formal, proper procedures for handling instances or accusations of misconduct. The latter must include protection for *both accuser and accused*, due process for both, and adjudicatory procedures that have credibility.

Neither academic institutions, nor laboratories (government or industry) have special qualifications for setting up such a system of procedures for fact finding, investigation, or adjudication. Especially where procedures may involve tangible or career penalties or may affect reputations, and where due process is called for, legal assistance may be necessary in setting up adequate mechanisms. Universities particularly are beginning to pick up this burden and to accumulate experience in handling cases including protections for whistle blowers. Ways of sharing that experience and word about which procedures are fair and effective are needed. So far, the ways of sharing information and experience appear to be inadequate.

To fill this gap, the panel proposed establishment of a privately-funded Science Integrity Advisory Board (SIAB). The board would be operated outside the scientific community and funded by foundations or other sources with no vested interests. The board would aid institutions on request, and would collect data and information from all sources to assess the state and trajectory of research ethics in the country. The panel believed this mechanism would operate without infringing on the leeway for innovative research techniques and thinking, which are often harbingers of revolutionary discoveries and technologies. The board as proposed would have a five-year sunset provision in its charter.

Our panel took a strong stand on formal guidelines for research practice. It said that such guidelines may be useful but they should be a local responsibility, not a list handed down by government research sponsors, or administrators. If guidelines were handed down, they would likely reside in desk drawers, at best. If the people who must live by guidelines participate in drawing them up, they are more likely to take note, and to make rules suitable for the research requirements of the specific disciplines involved.

This aspect of the panel's report has been criticized as being too weak. There are people, some in the scientific community itself, who take an authoritarian stance. They would like to see a more rigid and uniform governance of research. The panel believes such an approach would be less effective than our proposal in sustaining and improving research integrity. To aid institutions and laboratories in drawing up their own guidelines, the panel included in its report a minimal list of issues to be addressed by research guidelines.

VII. Conclusion

In closing, let me say that the panel's report has not been widely acclaimed or embraced. Opinion in the community and among the sponsors of research is quite diffuse. Some people believe the subject has been overplayed by the media and political forces; some believe that it is underplayed and the report is not severe enough. The idea of central authority dies slowly, as does the laissez-faire idea. The panel's report is between these extremes. Fortunately, many institutions, individuals, and laboratories are proceeding down this intermediate path.

The report did not address several vital issues related to misconduct in science. Prominent among these is conflict of interest. This subject deserves a study on its own since the research community is becoming increasingly involved in matters where economic rewards can be large, and where political advocacy can yield power and influence as well as research funding, psychic income, and ego satisfaction. The panel was not asked to examine the conflict issue except as it might affect the integrity of the research process. In that regard, the panel said that disclosure of potential conflicts is the best remedy.

Finally, let me say that I am personally encouraged by Congressman Dingell's statement, and I quote "The report represents a sea change in the scientific establishment's approach to issues of misconduct." My own addendum is that the establishment must now move forward resolutely down this path. Lacking that, the scientific community's traditional self-governance will be increasingly in jeopardy. The panel believed that sustaining that mode of governance is essential for a productive research system.

Speakers' Vitae

John F. Ahearne is executive director of Sigma Xi. He earned a B.S. in engineering physics and an M.S. in physics at Cornell University and an M.A. and Ph.D. in physics from Princeton University. From 1959 to 1970 he served in the United States Air Force. From 1972 to 1977 he served as Deputy and then Principal Deputy Assistant Secretary of Defense. He worked at the White House Energy Office and served as a Deputy Assistant Secretary of Energy from 1977 to 1978. From 1978 to 1983 he was a Commissioner of the U.S. Nuclear Regulatory Commission (Chairman from 1979 to 1981) and from 1984 to 1989 he was a vice president and senior fellow at Resources for the Future. He has served on numerous committees, including his recent work as chairman of the Secretary of Energy's Advisory Committee on Nuclear Facilities Safety.

Lisa Backus graduated summa cum laude from Harvard University, where she received a Bachelor of Arts in Psychology and Social Relations (Psycho-biology option) and was selected as a Rhodes Scholar. She received a D.Phil. in Clinical Pharmacology from St. John's College, Oxford University. Her thesis was "Functional interactions between 5-HT receptor subtypes: implications for the actions of psychotropic drugs." She currently is a postdoctoral fellow in the Program in Medical Ethics at the University of California, San Francisco, Medical School, from which she will receive her M.D. in May 1994. She is doing research on ethical issues in biomedical research including authorship, conflicts of interest, and intellectual property. Dr. Backus was a John Harvard Scholar, Elizabeth Cary Agassiz Scholar, and Detur Scholar at Harvard and was elected to Phi Beta Kappa.

J. Michael Bishop teaches and studies microbiology at the University of California, San Francisco, and is an internationally recognized authority on the molecular mechanisms of cancer. A graduate of Gettysburg College, he received his M.D. at Harvard University, followed by two years of training in internal medicine at Massachusetts General Hospital and another two years in the Research Associates Program at the National Institutes of Health. He then joined the faculty at the University of California and began his work on cancer. Dr. Bishop directs the G.W. Hooper Research Foundation at the University of California, San Francisco, where he is Professor of Microbiology, Immunology, Biochemistry, and Biophysics. A member of the National Academy of Sciences, Dr. Bishop is the recipient of many honors, including the 1989 Nobel Prize in Physiology or Medicine, with his colleague Harold Varmus, for their discovery that normal cells contain genes capable of becoming cancer genes.

George Bugliarello, President of Polytechnic University since 1973, is an engineer and educator with a broad background ranging from civil engineering to computer languages, biomedical engineering, and fluids mechanics. He is president of Sigma Xi. Dr. Bugliarello holds a Doctor of Science degree in engineering from the Massachusetts Institute of Technology, as well as honorary degrees from Carnegie-Mellon University, the University of Trieste, and the Milwaukee School of Engineering. He has served as Chair of the Board of Science and Technology for International Development (BOSTID) of the National Academy of Sciences and is Past-Chairman of the National Science Foundation's Advisory Committee for Science and Engineering Education. A past president of the Association of Independent Technological Universities and of the National Association for Science, Technology and Society, Dr. Bugliarello is founder and editor of *Technology in Society — An International Journal*, has authored some 200 professional papers, and is the author or co-author of numerous books. He is a member of the National Academy of Engineering.

Rita R. Colwell is Professor of Microbiology, President of the Maryland Biotechnology Institute, and former Director of the Center of Marine Biotechnology and of the Sea Grant College — all at the University of Maryland. She is immediate Past-President of Sigma Xi. Dr. Colwell is a microbiologist and marine scientist who is active in national and international research and teaching. The author of 12 books and approximately 400 scientific publications, she has served on the National Science Board. The recipient of several honorary degrees, she has been a Guest Scientist at the National Research Council of Canada and has served as Vice President for Academic Affairs for the University of Maryland System. A Past-President of the American Society of Microbiology, Dr. Colwell is a Fellow of the American Association for the Advancement of Science. She is also Chair of the Board of Governors of the American Academy of Microbiology and President of the International Union of Microbiological Societies. Among the awards she has received are the International Institute of Biotechnology Gold Medal and the Purkinje Medal for Achievement in the Biological Sciences from the Czechoslovakian Academy of Sciences. In 1992, she was named a National Scholar by Phi Kappa Phi.

Edward E. David, Jr. is President of EED, Inc., advisors to industry, government, and universities on technology, research, and innovation management. He chaired the National Research Council Panel on Scientific Responsibility and the Conduct of Research that published *Responsible Science: Ensuring the Integrity of the Research Process*, otherwise known as "The David Report." During his career, Dr. David has been Science Advisor to the President of the United States and Director of the White House Office of Science and Technology, President of Exxon Research and Engineering Company, and Executive Director of Bell Telephone Laboratories. He is the U.S. Representative to the NATO Science Committee. A former president of the American Association for the Advancement of Science, Dr. David is a member of the National Academy of Engineering, the National Academy of Sciences, the National Academy of Public Administration, and the American Philosophical Society. He is a trustee of the John S. Guggenheim Foundation. Dr. David received his Ph.D. in electrical engineering from the Massachusetts Institute of Technology. The recipient of 12 honorary degrees, he has received many awards, including the MIT Silver Stein Award, the Delmer S. Fahrney Medal from the Franklin Institute, the Arthur M. Bueche Award from the National Academy of Engineering, and the Industrial Research Institute Medal.

Carl Djerassi is Professor of Chemistry at Stanford University. Father of the birth control pill, developer of antihistamines, founder of chemical companies, and best-selling novelist, he is the winner of the 1992 Priestley Medal, the highest American award in chemistry, and received the National Medal of Science in 1973. In 1991, President George Bush presented him with the National Medal of Technology for his initiatives in developing new approaches to insect control. Dr. Djerassi received his A.B. from Kenyon College and his Ph.D. from the University of Wisconsin. Prior to joining the faculty at Stanford, he was a professor at Wayne State University. A former president of Syntex Research, he helped found Zoecon Corporation, now a subsidiary of Sandoz, Ltd., serving as Chief Executive Officer and Chairman of the Board. Dr. Djerassi has published more than 1,000 articles and seven books dealing with the chemistry of natural products, as well as poetry and short story collections, the novel *Cantor's Dilemma*, and the critically-acclaimed autobiography *The Pill, Pygmy Chimps*, and *Degas' Horse*.

Bernard Gert is Eunice and Julian Cohen Professor for the Study of Ethics and Human Values at Dartmouth College and Adjunct Professor of Psychiatry at Dartmouth Medical School. He helped found the Institute for the Study of Applied and Professional Ethics at Dartmouth. Dr. Gert is the author of *The Moral Rules: A New Rational Foundation for Morality*; co-author, with Charles M. Culver, of *Philosophy in Medicine: Conceptual and Ethical Issues In Medicine and Psychiatry*; and the author of *Morality: A New Justification of the Moral Rules (1988)*. The author or co-author of more than two dozen articles in medical ethics, he is currently the principal investigator on a grant from the National Institutes of Health to apply his moral theory to the ethical issues arising from the Human Genome Project, and co-investigator on a grant from the National Science Foundation for the production of educational modules for the teaching of research ethics. He was the only philosopher member of the Panel on Scientific Responsibility and the Conduct of Research, of the Committee on Science, Engineering, and Public Policy (COSEPUP) of the National Academy of Sciences, National Academy of Engineering, and the Institute of Medicine, which published *Responsible Science: Ensuring the Integrity of the Research Process*.

David L. Goodstein is Vice Provost and Professor of Physics and Applied Physics at the California Institute of Technology. A graduate of Brooklyn College who received his Ph.D. from the University of Washington, he is the author of more than 100 research articles and a book, *States of Matter*. His area of interest is condensed matter physics, especially two-dimensional matter and phase transitions. Dr. Goodstein has served on many scientific and academic panels, including the Standing Review Board of the Keck Telescope and the Jet Propulsion Laboratory Advisory Council. In the early 1980s, Dr. Goodstein was the host and project director for *The Mechanical Universe*, a 52-part college physics telecourse based on his popular lectures delivered over a three-year period at Caltech. The project, which has been adapted for high school use as well, has been broadcast on some 100 public broadcasting stations and has garnered more than a dozen prestigious awards, including the 1987 Japan prize for television.

Jan Gurley received a Bachelor of Arts degree in English from Emory University and was a Robert T. Jones Scholar at St. Andrews University, St. Andrews, Scotland. She received her M.D., magna cum laude, from Harvard Medical School and served her residency in General Internal Medicine at the University of California, San Francisco (UCSF). She currently is a postdoctoral Robert Wood Johnson Clinical Scholar at UCSF. Dr. Gurley has served on the Residency Review Committee, UCSF and is Co-chairman of the San Francisco Interns and Resident's Association. Dr. Gurley received the Henry Asbury Christian Award for notable scholarship in research at Harvard Medical School and was an Albert Schweitzer Fellow at the Schweitzer Hospital, Gabon, Africa. Dr. Gurley was elected to Phi Beta Kappa and was a member of the All-Scotland National Basketball Team and on the varsity basketball team at St. Andrews University.

Gerald Holton is Mallinckrodt Professor of Physics and Professor of the History of Science at Harvard University. Since 1976, he has also been Visiting Professor at the Massachusetts Institute of Technology as a founding member of the Science, Technology and Society Program. His books include *Scientific Imagination: Case Studies*; *Thematic Origins of Scientific Thought: Kepler to Einstein*; *The Advancement of Science, and its Burdens: The Jefferson Lecture and Other Essays*; *Albert Einstein, Historical and Cultural Perspectives*; and *Introduction to Concepts and Theories in Physical Science*. Dr. Holton is a member of the Editorial Committee of *The Collected Papers of Albert Einstein* being published by Princeton University Press. He was the founding editor of the quarterly journal *Daedalus*, and founder of the journal *Science, Technology, and Human Values*. His many honors include the George Sarton Medal, the J.D. Bernal Prize, the Gemant Award, the Millikan Medal, the Oersted Medal, a Guggenhein Fellowship, and selection by the National Endowment for the Humanities in 1981 as the tenth annual Jefferson Lecturer.

Garth Jones began studying chemistry at Grinnell College, and graduated from that institution in 1988. During his undergraduate years he learned the fundamentals of research from Professor Luther Erickson at Grinnell and, during the summers, from Professor Louis Messerle at the University of Iowa. He completed a Ph.D. in chemistry in 1992 at the University of Iowa under the direction of Professor William Scott. His thesis focused on studying the oxidative-addition of palladium(0) catalysts to carbohydrate electrophiles and the reaction chemistry of 2-alkoxyglycals. In the fall of 1992 he moved to Stanford University where he is currently a postdoctoral associate in the laboratories of Professor William S. Johnson.

Yuan T. Lee, University Professor and principal investigator for the Lawrence Berkeley Laboratory at the University of California, Berkeley, received his B.S. degree from National Taiwan University, his M.S. from Tsinghua University, and his Ph.D. from the University of California, Berkeley. Following a fellowship at Harvard University and a faculty appointment at the University of Chicago, Dr. Lee returned to Berkeley as a full professor and significantly expanded his research. In addition to crossed molecular beams studies of reaction dynamics, his interests include investigations of various primary photochemical processes and the spectroscopy of ionic and molecular clusters. Among his many honors and awards are the 1986 Nobel Prize in Chemistry and the 1986 National Medal of Science. Dr. Lee is a fellow of the American Academy of Arts and Sciences, a member of the National Academy of Sciences, a member of the Academia Sinica in Taiwan, a foreign member of the Göttingen Academy of Sciences, and a member of the Third World Academy of Sciences, among others. Dr. Lee is also a former Guggenheim Fellow.

Morris A. Levin received a Bachelor of Arts in Biochemistry from the University of Chicago and a Ph.D. in Bacteriology and Biophysics from the University of Rhode Island. He is a senior research scientist at the University of Maryland's Maryland Biotechnology Institute. He was coordinator for biotechnology risk assessment with the U.S. Environmental Protection Agency, has served as science advisor to the U.S. House of Representatives' Science and Technology Committee, and currently teaches microbial ecology at the National Institutes of Health's Foundation for the Advancement of Education in the Sciences.

Bernard Lo is Associate Professor of Medicine and Director of the Program in Medical Ethics at the University of California, San Francisco, and Co-Director of the UCSF-Stanford Robert Wood Johnson Clinical Scholars Program. He received his undergraduate education at Harvard University and his medical degree from Stanford University. He serves on the ethics committee of the American College of Physicians and on the editorial board of the *Annals of Internal Medicine*. The Program in Medical Ethics conducts seminars on ethical issues in the conduct of biomedical and clinical research for trainees at UCSF.

Thomas F. Malone has held posts as MIT Professor; Senior Vice President and Director of Research for the Travelers Insurance Companies; Dean of the Graduate School, University of Connecticut; Foreign Secretary of the National Academy of Sciences; Executive Scientist of the Connecticut Academy of Science and Engineering; President of the American Meteorological Society, of the American Geophysical Union, and of Sigma Xi; and Secretary General of the Scientific Committee on Problems of the Environment of the International

Council of Scientific Unions. In 1984, he received the World Meteorological Organization Prize for work done in the field of international meteorological organizations. In 1991, he was selected by a jury of 12 scientists from eight countries to receive the St. Francis of Assisi International Environmental Prize for his work as "initiator of major international and interdisciplinary environment research programs." He is currently also a Distinguished University Scholar at North Carolina State University and Director, The Sigma Xi Center in Research Triangle Park, NC.

Tara Meyer graduated in 1986 from Grinnell College with a B.A. in chemistry. She received her Ph.D. degree in inorganic chemistry in 1991 from the University of Iowa. Dr. Meyer's thesis work concerned the synthesis and reactivity of early transition metal compounds. After graduation, she spent one year as a postdoctoral associate under her thesis advisor, Professor Louis Messerle, and Professor Rich Jordan. In 1992, Dr. Meyer assumed her current position at the University of California, Berkeley where she is a joint postdoctoral associate with Professor Bob Bergman and Professor Bruce Novak.

Arthur L. Singer, Jr. is Vice President of the Alfred P. Sloan Foundation in New York. A graduate of Williams College, he received his M.B.A. at the University of Michigan. A former administrator at the Massachusetts Institute of Technology, an executive with the Carnegie Corporation, and President of the Education Development Center, Mr. Singer is a consultant in the field of higher education, a member of the visiting committee of the MIT Center for International Studies, The New School for Social Research's School of Management and Urban Policy, and the publications committee for The Public Interest.

Steven Weinberg is Josey Regental Professor of Science at the University of Texas at Austin. Well known for his research on the unification of elementary particle forces and for other major contributions to physics and cosmology, his work has been honored with 11 honorary doctoral degrees and many prizes and medals, including the 1979 Nobel Prize for Physics and the National Medal of Science. His prize-winning book, *The First Three Minutes*, has been translated into 20 foreign languages. Dr. Weinberg is also the author of *The Discovery of Subatomic Particles* and the influential treatise *Gravitation and Cosmology*, together with over 200 articles on elementary particle physics and cosmology. A member of the Royal Society of London, as well as the U.S. National Academy of Sciences, the American Philosophical Society, and the American Academy of Arts and Sciences, among others, he is an active participant in the Superconducting Super Collider project, having served on its Board of Overseers, Site Selection and Evaluation Committee, and Scientific Advisory Committee.

Rosalyn S. Yalow is a Senior Medical Investigator at the Bronx Veterans Administration Medical Center. She is also Solomon A. Berson Distinguished Professor-At-Large of the Mount Sinai School of Medicine, City University of New York, and Emeritus Distinguished Professor-At-Large of the Albert

Einstein College of Medicine, Yeshiva University of New York. In 1977, she was awarded the Nobel Prize in Physiology or Medicine for the development of radioimmunoassay, a methodology now used in thousands of laboratories around the world to measure hundreds of substances of biologic interest in blood and other body fluids. In 1988, Dr. Yalow received the National Medal of Science from President Ronald Reagan. A member of the National Academy of Sciences and of the American Academy of Arts and Sciences, she is the recipient of 51 honorary degrees and many awards, including the first Veterans Administration William S. Middleton Award for Medical Research, the Albert Lasker Basic Medical Research Award, and the Gairdner Foundation International Award. Dr. Yalow has served as President of the Endocrine Society and as a member of numerous advisory agencies and editorial boards.

Appendix A

Breakout Group
Conclusions and Recommendations

Breakout groups afforded Forum participants an opportunity to discuss in greater depth topics related to ethics and research. Each group met twice, with the first session begun by presentations by a panel. (Some panelists prepared papers. These are included in Appendix B.) The breakout groups were charged with developing no more than three recommendations and three conclusions. The following were developed by the participants, and do not necessarily represent the views of the moderators, panelists, or of Sigma Xi.

The Peer Review Process

Moderator: **Drummond Rennie**, Deputy Editor, *The Journal of the American Medical Association*

Panelists: **David L. Goodstein**, Vice Provost, California Institute of Technology

R.E. Sojka, U.S. Agricultural Research Service, Kimberly, Idaho

Rapporteur: **Jackie B. Langston**, Programs Administrator, Sigma Xi

The peer review process is used to evaluate the publication of articles, allocate funds to projects and determine personnel decisions. We focused primarily on the editorial peer review process.

Conclusions

1. We should retain peer review because it serves science better than any alternative.

2. Nevertheless, we perceive it to have severe problems that need to be investigated and remedied.

3. Differences exist among disciplines, but certain universal principles should be applied to all.

Recommendations

1. Authors and their institutions should never be identified to referees.

2. Editors should:
   - take full responsibility for final publishing decision
   - ensure timely handling of manuscripts
   - use more than one referee per manuscript and be the final arbiter and judge of disputes between referees and authors
   - acknowledge referees
   - not hide behind referees
   - send all referee comments to authors and other referees
   - be responsible for confidentiality of the system

3. Sigma Xi should sponsor an investigation of the efficacy of the peer reviewed system (to gather information).

Institutional Responses to Misconduct in Science

Moderator: **Nicholas Steneck**, Professor of History, University of Michigan, and Chairman, Advisory Committee on Scientific Integrity, U.S. Public Health Service

Panelists: **Steven M. Blush**, Director, Office of Nuclear Safety, U.S. Department of Energy

Jerome L. Rosenberg, Research Integrity Officer, University of Pittsburgh

James J. Zwolenik, Assistant Inspector General for Oversight, National Science Foundation

Rapporteur: **G. Dale Buchanan**, Professor of Biomedical Science, McMaster University

Conclusions

1. While institutions have demonstrated improvement in responding to allegations of misconduct, continued attention should be directed toward how to respond to and deal with misconduct allegations.

2. Misconduct is often an outgrowth of mismanagement or lack of supervision.

3. The lack of institutional memory and experience often inhibits proper investigation of allegations of misconduct.

Recommendations

1. That organizations such as Sigma Xi, the American Association of Medical Colleges, American Association of Universities and others continue their good offices in encouraging institutions to respond to misconduct allegations in a constructive way and, if possible, serve as a reference base in obtaining expert assistance and guidance.

2. In responding to allegations of misconduct, institutions should not only examine what occurred, but seek to ascertain why or how it occurred.

3. There should be a designated authority who will serve as this institutional "memory" and locus of information and as a focal point.

Improving Mentoring in the Academic Research Environment

Moderator: **R. E. Fornes**, Professor of Physics and Associate Dean of Physical and Mathematical Science, North Carolina State University

Panelists: **Carl Djerassi**, Professor of Chemistry, Stanford University, and former President, Syntex Research

Neal Lane, Provost, Rice University

Rapporteur: **Geraldine Twitty**, Professor of Zoology, Howard University

The importance of mentoring in the academic research environment, in the advancement of science, and in the development of scholars in the field is widely recognized. However—

Conclusions

1. Significant problems exist in mentor/mentored relationships and the problems are widespread.

2. Issues are field, gender, rank, and ethnic-dependent.

3. Major reluctance exists among mentors, mentored, and institutions to address the problem.

Recommendations

1. Establish an *ad hoc* committee of Sigma Xi to first define the ideal role of mentor/mentored relationships in the academic research environment and to assess the problems of mentoring. This should draw on existing reports, publications, workshops, and perhaps additional information gathered by a possible questionnaire.

2. The *ad hoc* committee should develop a set of guidelines to address problems of mentoring.

3. The committee should develop recommendations for the effective implementation of a plan to resolve problems.

Definitions of Misconduct in Science

Moderator: **Tomuo Hoshiko**, Professor of Physiology and Biophysics, Case Western Reserve University

Panelists: **Donald E. Buzzelli**, Senior Scientist, Office of Oversight, Office of the Inspector General, National Science Foundation

            Howard K. Schachman, Professor of Molecular and Cell Biology, University of California, Berkeley

Rapporteur: **Peggy Wayne**, Professor of Biology, Oral Roberts University

No agreement was reached regarding inclusion of the phrase "and other serious deviations from accepted practices."

Conclusions

1. The definition of misconduct in science is designed to determine which behavior is to be sanctionable by the scientific misconduct apparatus of federal agencies. These actions do not constitute the full range of unacceptable behavior by scientists.

2. Included in misconduct are fabrication, falsification, plagiarism, and other deliberate misrepresentations in proposing, performing, reporting, or reviewing research.

Recommendations

1. Scientific institutions and societies should generate more specific guidelines for responsible conduct in research.

Science and the Media

Moderator: **Alan McGowan**, President, Scientists' Institute for Public Information

Panelists: **Deborah Blum**, Pulitzer Prize-winning science writer, *The Sacramento Bee*

            David Perlman, Science Editor, *The San Francisco Chronicle*

Rapporteur: **Timothy Ng**, Professor of Horticulture, University of Maryland

Science and technology are in the media spotlight more than ever before in the past, and this will probably increase. The media has a tremendous potential to further the cause of science and technology.

Conclusions

1. Scientists and journalists are working towards a common cause — to inform the public.

2. Scientists and journalists have a mutual responsibility for accurate, open and balanced information.

3. Scientific issues can rapidly escalate into social, ethical, and political issues.

Recommendations

1. Increase the interaction between scientists and journalists to foster understanding through seminars and informal meetings and get-togethers.

2. Increase the flow and improve the clarity of information from scientists to journalists through workshops for scientists on how to interact with the media and by helping the media to identify credible sources of information.

3. Increase the ability of journalists to report accurately on this improved clarity and increased flow of information through workshops/briefings held by scientists for journalists and also by increasing the recognition of scientific journalism as an educational discipline.

Teaching Ethics

Moderator: **Joseph M. Norbeck**, Director, Center for Environmental Research & Technology, University of California, Riverside

Panelists: **Stephanie J. Bird**, Research Associate and Lecturer, Department of Brain and Cognitive Science, Massachusetts Institute of Technology

George Sammet, Jr., Vice President-Ethics, Martin Marietta Corporation

Vivian Weil, Director, Center for the Study of Ethics in the Professions, Illinois Institute of Technology

Rapporteur: **Michael LaBarbera**, Professor of Organismal Biology & Anatomy, University of Chicago

Conclusions

1. Ethical behavior of scientists and engineers is the primary responsibility of the individual, but also requires appropriate policy support of institutions and society and encompasses all aspects of one's professional life. Attention to ethical issues is in the long-term interests of both individuals and institutions.

2. Science and engineering have specific issues that need to be considered, including the everyday conduct of scientific work, the social organization of science and engineering, and the interaction of science with the rest of society.

3. Appropriate ethical behavior must be communicated to and practiced at all levels of academic, governmental, industrial, and other research organizations associated with science and engineering. It is possible and necessary to teach ethics in order to increase sensitivity and enhance the ability to recognize and analyze the ethical dimensions and issues of everyday scientific practice.

Recommendations

1. We strongly recommend the teaching of ethics for all students of science as an integral component of formal scientific education. In this connection, we urge that there be cooperation between technical professionals and scholars in the humanities.

2. We recommend that formal discussion, reevaluation, and continuing education about ethical issues be institutionalized in all scientific and engineering organizations.

3. We recommend that Sigma Xi establish ongoing programs to foster recognition of the importance of teaching ethics in science and engineering. Initiatives might include:

   (a) Prepare a brochure on ethical behavior in science and engineering

   (b) Annually recognize exemplary ethical behavior

   (c) Establish national and/or regional lectureships on ethical issues

   (d) Prepare and distribute case studies for use by local chapters or clubs

   (e) Foster interactions on ethical issues with other scientific organizations

The Ethics of Diversity

Moderator: **Peggie J. Hollingsworth**, Research Scientist, Department of Pharmacology, University of Michigan

Panelists: **Catherine Didion**, Executive Director, Association for Women In Science

Aaron Wildavsky,[1] Professor of Political Science and Public Policy, University of California, Berkeley

Rapporteur: **Charles Smith**, Professor of Pharmacology, University of Michigan

Conclusions

1. Despite the controversy as to whether there is a future shortage of scientists and engineers, there is absolutely no doubt that women, persons of color, the physically challenged and others from diverse cultures are inadequately represented in the community of scientists and engineers. This is an intolerable situation that calls for active intervention. Increased participation by such individuals in science and engineering is critical for the future well-being, vitality, viability, and public acceptance of the scientific enterprise. Increased diversity will enhance, energize, and bring new dimensions of excellence to the scientific enterprise.

2. Numerous and clearly evident barriers exist for women, persons of color, and others from underrepresented groups that prevent upward mobility and deny equal access to positions of authority and power. The persistence of such barriers will have a negative impact upon the achievement of diversity in science and engineering.

3. Increased inclusion of women, persons of color and others from underrepresented groups in science and engineering need not, and must not, adversely affect the high standards of excellence that characterize the modern scientific enterprise.

Recommendations

1. Members of the scientific community must continue and expand upon their efforts to educate the public and attract to careers in science women, persons of color and others from underrepresented groups. These efforts should start with children of all ages, their families and teachers, and must continue undiminished through college undergraduate, graduate-school, and professional-school years.

---

[1]Dr. Wildavsky notes that he was unable to attend the session during which the conclusions were formulated and disagrees entirely.

2. The scientific community must take immediate, direct and forceful actions to remove those barriers to access and to upward mobility commonly encountered by women, persons of color and others from underrepresented groups who have demonstrated ability and potential, who have sought out and completed advanced training with distinction, and who have shown scientific productivity. A partial list of such barriers include lack of mentoring, uneven granting of research funding at the federal, state and local levels, inappropriate and excessive assignments to committees and to low-level administrative and teaching duties, unequal financial remuneration, delayed advancement and promotion, and negative ethnic and gender stereotyping.

3. The scientific community must be open and receptive to new ideas and novel approaches to science that will inevitably accompany increased diversity. At the same time care must be taken to instill in all who join in the scientific enterprise a proper appreciation of traditional standards of excellence and scientific integrity.

The Societal Responsibilities of Science

Moderator: **John Bailar**, Professor of Epidemiology and Biostatistics, McGill University

Panelists: **Gerald Holton**, Mallinckrodt Professor of Physics and Professor of the History of Science, Harvard University

Chauncey Starr, President Emeritus, The Electric Power Research Institute

Rapporteur: **Timothy Weiskel**, Director, Harvard Seminar on Environmental Values, Harvard University

Scientists form a dynamic element of human society, reflecting its diversity, its values, and its aspirations. In the post World War II era an implicit social contract emerged between science and society involving the progressive application of scientific discoveries to the achievement of broadly perceived social goals. In our day as well, many of the critical issues facing society beg for improved scientific information and understanding, but circumstances have changed. Today it is recognized that the connection between issues is more complex and profound than it was thought to be in the immediate post-war era. In addition, the alienation of the public toward science has made the process of bringing scientific understanding to the public and policy-makers more difficult.

200

Conclusions

1. The events prior to and during World War II led to an extraordinary marriage of science and public policy which greatly accelerated the development of certain scientific disciplines. It also led to an enormous increase in the public's expectations of science. Many of these expectations in the social, economic, and political realms are bound to be disappointed.

2. Scientific research has been so remarkably successful in the post-war era because of its application of a reductionist approach to problems and the development of solutions by specialized scientists. In our day, however, the application of science to the solution of contemporary complex and interconnected problems requires a more interdisciplinary approach both by individual scientists and groups of scientists.

3. An awareness of societal consequences will require that scientists increase their educational efforts to convey an understanding of their work to the lay public, to participate more directly in the political process, and to exercise the highest ethical standards in their work. These efforts will help to build and maintain the respect and support of the public for the scientific endeavor.

Recommendations

1. As an interdisciplinary, grassroots, comprehensive organization of scientists, Sigma Xi should initiate two-way communication processes with the public and policy-makers on behalf of the scientific community. Specifically, Sigma Xi and its membership should initiate an examination of the process by which science is brought to bear in policy-making by the legislative and executive branches of government.

2. Sigma Xi should devise public information programs to educate the public about the scientific method and the legitimate expectations and limitations of science.

3. Scientists have the responsibility to be aware of the social consequences of their activities, and to participate in resolution of related issues. Science must fully disclose to the public its capabilities, its limitations, and the importance of its participation in addressing today's social and ethical problems.

Appendix B

Panel Presentations

Some panelists in the breakout group sessions (Appendix A) provided written comments. None were asked to and most did not. Therefore, the following is only a sample of what attendees at the breakout sessions heard. However, readers of the proceedings will benefit from these papers. In addition, Judith Swazey provided a background paper for one panel. This paper is included.

Presented in the breakout session on The Peer Review Process.

Driving Science With One Eye on the Peer Review Mirror

R. E. Sojka and H. F. Mayland

We would like to share some insights from recent experiences in defining the ethical and procedural framework governing peer-reviewed scientific publishing. These insights ultimately point to the need for greater "enfranchisement" [participatory democracy] for scientific authors.

"Publish or Perish" is a phrase that may have originated with Kimball C. Atwood, then of Columbia University, sometime during or shortly before 1950[1]. It is the scientists' equivalent of "The buck stops here; Cherche' la femme; and Live long and prosper" all rolled into one maxim. And yet the story of its utterance is a lesson unto itself. For today its origin with Atwood can only be documented anecdotally. Atwood never published the phrase, and as the story goes, had only to wait a month before he heard it delivered in an address by a visiting lecturer, who afterward told Atwood he heard the phrase from a participant in Atwood's originating conversation.

Publication documents the precedence of ideas. It documents the stewardship of research funds. It documents the productivity of scientists, justifies our salaries and our reputations, and allows the cultivation of our egos. But most importantly, it liberates information and knowledge from the imprisonment of chaos and file cabinets to the free access of other scientists and for the betterment of mankind. The publication ethic was being evoked as early as the mid-18th century by Benjamin Franklin, who exhorted scientists simply "To study, to finish, to publish."[2]

But, writing does not come easily to most scientists. Writing is certainly less gratifying than doing experiments, and it is certainly anticlimactic compared to the epiphanal experience of watching data being plotted for the first time on a computer screen. And as the protocols of refereed scientific publishing have evolved over its brief two hundred or so years of formal history, many scientists have developed significant concerns over the capriciousness of the process itself.

We have been immersed in the topic of scientific ethics and peer reviewing for over five years. In 1990 we organized a symposium on "Research Ethics, Manuscript Review and Journal Quality." The symposium has had a lasting impact on these issues within our 13,000-member American Society of Agronomy, Crop Science Society of America, Soil Science Society of America, known collectively as the Tri-Societies.

---

[1]Personal communication R.C. von Borstel, 30 October 1991.
[2]Mackay, A.L. 1992. *A Dictionary of Scientific Quotations*. IOP Publishing Ltd., Bristol and Philadelphia. p. 94.

Also, in the fall of 1989 one of us (R.E.S.) was appointed to the U.S. Department of Agriculture, Agricultural Research Service's (ARS) agency-wide "Committee on Misconduct in Science"; and over the course of three years that committee has developed a "Code of Scientific Ethics" for ARS's 2500 scientists and established the formal protocols for dealing with allegations or evidence of scientific misconduct. The committee also proposed a variety of measures for reinforcing the ethical climate within ARS's scientific work force.

From these involvements we have come to realize that "Peer Review" also bears a pivotal relationship to the entire spectrum of related ethics and misconduct issues. And while we will focus our remarks on peer review vis-à-vis the refereed journal publication process, the insights are essentially the same with reference to other uses of peer review; for example, review of funding proposals or project plans and personnel promotion and tenure evaluation.

The merits and pitfalls of scientific peer review can be roughly segregated into two broad categories. One encompasses what might be regarded as "the art of scientific communication." It includes the procedural idiosyncracies of journal editorial and review policy, the concern for effective presentation, and even the quality and accuracy of data interpretation. We would also include simple courtesy and reviewer/author professionalism and competency.

The second broad category encompasses what might be regarded as the "ethics of scientific communication." This category of merits and pitfalls is involved with recognizing or obscuring conflict of interest. A simple analogy might be worth stating at this point. If integrity is the glue of science, then conflict of interest is the solvent. The promotion of integrity and the curbing of conflict of interest define the essential ethics of scientific communication.

We should clarify our concern about publication ethics. The ethical framework we are attempting to define is the one related to the process of communicating, and is essentially separate from the content being communicated. We are deliberately sidestepping the issue of defining what is ethical or unethical research. That topic is undoubtedly a valid concern, but most would probably agree that it is an issue separate from the ethics of the peer review process per se.

The information data base used in developing this ethical framework was derived from both the symposium and from experience on the ARS Committee on Misconduct in Science. Symposium papers were compiled in a special publication.[3] It contains a fine collection of articles that touch on many aspects of the same issues considered in the Sigma Xi Forum held in San Francisco.

The symposium provided information on journal stature, on government interference with scientific freedoms, administration of ethical standards in the ARS, scientists' perceptions of the peer-reviewing and editing process, and the history of peer reviewing and editing. We feel that one of the most important

---

[3]Mayland, H. F. and R. E. Sojka. 1992. *Research Ethics, Manuscript Review and Journal Quality*. ACS Miscellaneous Publication. American Society of Agronomy, Madison, Wisconsin, Publishers.

things we did in the symposium was to ask questions and not formulate our opinions from personal supposition. This came in the form of 1) soliciting quantitative information about our profession's core journals, 2) analyzing the demographics of our membership related to involvement in the peer review process, and 3) conducting a scientific survey of essential peer-review-related issues within a balanced sample of our Tri-Society's membership.

Information gathered from the surveys and from papers presented at the symposium, plus interactions with other scientists was then summarized in a report which we prepared for the Tri-Society's governing board. The report included a series of recommendations relating to the peer review process of technical papers which we felt relevant to the Tri-Societies. The recommendations, which are shown below, are also germane to the Sigma Xi forum on "Ethics, Values, and the Promise of Science" and to other scientific societies.

1. Strive for demographic balance among editorial boards and reviewers.
2. Promote robust institutional review before journal submission to improve manuscript quality and possibly discourage misconduct.
3. Adopt dual anonymity (author and reviewer) for peer reviews.
4. Proactively cultivate author recognition of his/her primal responsibility for accuracy and quality of published manuscripts.
5. Investigate vehicles to limit journal liability for publication of flawed, inaccurate or fraudulent manuscripts.
6. Select competent reviewers and allow authors fair opportunity for critique and rebuttal of reviewer remarks.
7. Provide reviewer training and feedback to reviewers.
8. Recognize the publication validity of neutral and negative results, and of unconventional, innovative ideas.
9. Be vigilant against the bias of influence networks; i.e., bias resulting from political, academic, geographic, or technical interests.
10. Promote university requirements for training in technical writing, reviewing, and editing that simulates target journals.
11. Promote institutional and professional codes of scientific ethics.

The most important product of the ARS agency-wide Committee on Misconduct in Science was publication of USDA-ARS Directive 129.0 "Procedures for Reporting and Dealing with Possible Misconduct in Science," which contains an agency code of scientific ethics for use in judging conduct. In addition, a list of general recommendations was given to the ARS administrator on ways to enhance the ethical climate of the ARS.

That code of scientific ethics follows. You will recognize that many of its elements impact upon or are impacted by the peer-reviewing, editing, and publication process:

Code of Scientific Ethics

for the

United States Department of Agriculture,

Agricultural Research Service

*I dedicate myself to the pursuit and promotion of beneficial
scientific investigation, consistent with the
mission of the Agricultural Research Service.*

I will never hinder the beneficial research of others.

*I will conduct, discuss, manage, judge and report science
honestly, thoroughly, and without conflict of interest.*

*I will encourage constructive critique of my personal science and
that of my colleagues, in a manner that
fosters harmony and quality amid scientific debate.*

*I recognize past and present contributors to my science and will not
accept unwarranted credit for the accomplishments of others.*

*I will maintain and improve my professional skills
and be a mentor to others.*

*I will ensure safety and humane treatment of human and animal subjects
and will prevent abuse of research resources entrusted to me.*

We have also promoted a fuller enfranchisement of scientists within their
scientific societies and within the journals in which they publish.

Of course, enfranchisement in the peer reviewing and editing process
demands a balancing of rights and responsibilities, but neither of these can be
dictated. They must be arrived at through a process of consensus determination
among participating scientists and the institutions which support them. And
this is best done within the framework of individual scientific professional
societies, especially those that manage the journals in which we publish. That
is not to say, however, that a broader consensus should not in fact be achieved
through coalitions of societies and broad-based philosophically disposed societies such as Sigma Xi.

In this case, enfranchisement means having a measure of say-so regarding how the vehicle of scientific publication is driven. Our administrators,
professional society officers, and editors are entrusted to place their hands on
the wheel, feet on the pedals, and eyes forward on the road ahead; but hopefully they also cast an occasional prudent glance in the rear view mirror, not
just to see where we have been, but also to see what 18-wheeler may be

gaining on us; and hopefully they cast another occasional glance at their passengers and ask how they are doing. For although every passenger in that vehicle cannot steer and brake to their personal satisfaction, neither should obdurate drivers have the power or freedom to take the passengers where they do not want to go, nor should they be allowed to take them there in unnecessary discomfort.

Abstract

Research reports are generally critiqued by fellow scientists. The action, otherwise known as the peer-review process, is subject to various abuses. This paper draws on the authors' experience with the peer-review process, on information presented at a 1991 symposium on the topic, and on a survey of perceptions by members of the American Society of Agronomy (ASA). The paper lists 11 recommendations made to the ASA board of directors. These include selection of competent reviewers and allowing authors fair opportunity for critique and rebuttal of reviewer remarks. Also included was a recommended promotion of institutional and professional codes of scientific ethics. The code for the Agricultural Research Service, which one of us helped develop, is provided as an example.

Minimizing Government Intervention in Scientific Institutions, and Regulating Misconduct in the Government Research Organization

Steven M. Blush[1] and H. T. Anderson[2]

I. Introduction

The U.S. Department of Energy (DOE) is responsible for managing the largest complex of research laboratories and scientific user facilities in the world. Allegations of scientific misconduct similar in many respects to the more publicized cases of misconduct that have been investigated by the National Science Foundation (NSF) and the Public Health Service (PHS) these past few years have also arisen at some of the DOE laboratories.

Unlike NSF or the PHS, DOE has not issued any regulations explicitly addressing the problem of misconduct in science. However, DOE is a participant in an interagency group that is meeting to develop a uniform federal government approach to handling allegations of misconduct in science, and DOE has a track record of investigating such allegations.

The Department has evolved a rather complicated framework for dealing with allegations of scientific misconduct. Three separate DOE headquarters offices may be called upon to investigate an allegation of misconduct: the Office of the Inspector General, the Office of Contractor Employee Protection and the Office of Nuclear Safety (ONS). The reason these three offices share jurisdiction in this area is because an allegation of misconduct could entail a violation of one or more of several different federal statutes, and the regulatory responsibility for these statutes is divided differently among the three offices. This promotes in particular offices within the agency the accumulation of specialized knowledge and expertise concerning particular statutory requirements. It also necessitates close coordination among the offices to avoid problems. In light of the possibility of duplicative investigations that would waste federal resources and potentially undermine due process protections, the Department has established specific mechanisms for sorting out which office should take the lead in a particular case.

The Office of Nuclear Safety has responsibility for investigating cases of alleged misconduct at DOE facilities involving suspected violations of Sections 223, 226 or 234 of the Atomic Energy Act. (These sections of the Act authorize civil and criminal penalties for tampering with restricted data or for violating DOE's nuclear safety related rules, regulations and orders.) ONS would also be

[1]Director of Nuclear Safety, U.S. Department of Energy
[2]Partner, Stier, Anderson & Malone, Washington, D.C.

assigned those cases that involve suspected violations of regulations covering fitness for duty and protection of whistle blowers at DOE nuclear facilities. Allegations of scientific misconduct would be investigated by ONS if, for example, nuclear safety related data were alleged to have been fabricated or falsified at a DOE laboratory, or if an employee claimed to have been retaliated against for alleging scientific misconduct involving the fabrication or falsification of nuclear safety related data. ONS will take the lead in these cases to minimize the possibility that an existing nuclear hazard could go overlooked during the course of an investigation into an allegation of misconduct involving nuclear safety related data.

This paper focuses on lessons the authors have learned about how scientific institutions can minimize government intervention in the investigation of allegations of misconduct, and presents an approach to regulating conduct in the special circumstances of the government research organization.

II. Investigating Misconduct Allegations in Scientific Institutions: How to Minimize the Likelihood of Government Intervention

Scientific institutions, like other organizations, respond to allegations of misconduct with varying degrees of effectiveness. One measure of effectiveness for institutions subject to governmental oversight is the extent to which the internal handling of the allegations instills confidence that the institution has resolved the issue. Such confidence often prevents government intervention or at least minimizes its scope. This section identifies some of the most common pitfalls in institutional responses to misconduct allegations — the kinds of mistakes that often create public whistle blowers and invite full-scale government intervention — and offers suggestions for avoiding these problems.

An investigation that will withstand government scrutiny nearly always requires the investigator to probe beyond the often narrow confines of the allegation itself and discover the root cause of the event or condition underlying the controversy. This process often leads the investigator far from the scientific discipline that is ostensibly the focus of the allegation and into realms of human and organizational behavior in which scientists possess no special expertise.

An allegation of falsification, for example, might begin innocently enough with one scientist, in the spirit of peer review, expressing concern about the validity of a colleague's data. If the colleague accepts the criticism in the spirit in which it is offered, he may correct the flaw in his data or, alternatively, demonstrate to the other scientist that the concern is unwarranted. If instead the colleague responds to the concern with an emotional, *ad hominem* attack on his fellow scientist's motives, what started as an informal exercise in peer review can degenerate into a personal feud between the two scientists that results in charges of misconduct. Timely intervention and sound judgment by the managers of the feuding scientists might prevent such escalation, but unresponsive or heavy-handed managers can aggravate the dispute. The root cause of a poor management response, in turn, might be traced to specific conditions pertaining to an individual manager (*e.g.,* alcoholism) or to an overall management system that, in the argot of root cause analysis, "sets up" individual managers to fail.

These human variables are just as important in scientific institutions as they are in businesses and other organizations. There is no *a priori* reason to believe that supervisors in scientific laboratories are less likely to be guilty of favoritism, discrimination, harassment, or a host of other management failings than their counterparts in business or government. Nor are scientists working together free of the jealousies, rivalries, miscommunications and other human foibles that give rise to an infinite variety of misconduct allegations. Even when such allegations are not on their face "scientific," they can become part of the root cause analysis conducted by those charged with investigating misconduct in science. Just as favoritism or a love triangle can, under some circumstances, seriously impede the effectiveness of a car dealership, such conditions can undermine the quality and efficiency of scientific work. Thus, a key issue when misconduct occurs in a scientific setting, as elsewhere, is the extent to which the misconduct affected or had the clear potential to affect the work of the organization.

Many of the same human foibles that produce misconduct allegations tend to prevent managers, investigators, and others in the organization from properly responding to them. Among the most common deficiencies in institutional responses to misconduct allegations are the following: failure to determine whether the investigation should focus on fact-finding, dispute resolution, or a combination of the two; failure to identify the standards of conduct at issue and to re-identify them as the evidence is gathered and the true nature of the incident unfolds; failure to analyze how the institution's management, peer review, and other systems functioned in response to the controversy giving rise to the investigation; and failure to integrate the technical and human behavior aspects of the investigation. As discussed below, how an institution responds in these four areas often determines the likelihood of government intervention.

A. Fact-Finding Versus Dispute Resolution

Allegations of misconduct that become public or attract the attention of a government agency are often rooted in personnel disputes that could have been resolved long before they created a "whistle blower" mentality that resulted in an open attack on the institution. Conversely, other whistle blowers have been created precisely because management treated their concerns solely as personnel issues, disregarding the safety, quality, or other substantive allegations embodied within those concerns. In both types of cases misconduct charges can be escalated unnecessarily, or an institution's response to serious misconduct be deemed inadequate, because of management's failure to grasp the distinction between fact-finding and dispute resolution.

Some controversies can be resolved satisfactorily with little or no fact-finding. In a matrimonial dispute, for example, the public has little compelling interest in determining who forgot whose birthday and the many other facts and circumstances that might explain the deterioration of a marital relationship. Provided the interests of any children or other third parties are properly accounted for, if the disputing parties are happy with a settlement of the conflict, the rest of us are usually happy as well.

On the other hand, if a dispute involves a matter of vital public interest, simply proposing a compromise settlement that makes the disputing parties happy will not be a satisfactory resolution. When the space shuttle *Challenger* blew up, for example, a potential dispute was created between the relatives of the dead astronauts and those responsible for the disaster. At the same time, there was a vital public interest in knowing, with great precision, what conditions led to the explosion. This required fact-finding, not dispute resolution. Settling the dispute with the astronauts' grieving relatives, no matter how satisfactorily, would not establish the root cause of the accident or prevent a future similar occurrence.

As the *Challenger* example illustrates, many controversies involve elements of both fact-finding and dispute resolution. Identifying the two kinds of issues at the earliest possible time and addressing each in an appropriate manner requires great skill and sensitivity on the part of an institution's management. Some charges of misconduct are made in the heat of emotions created by a personnel dispute that, if handled in a timely and sensitive manner, can be resolved quickly and with minimal damage to the individuals involved and little disruption to the work of the organization. It is a common mistake, however, to investigate and attempt to resolve both the personnel and substantive issues arising from a misconduct charge in the same way.

Personnel issues usually require elements of the "adversarial" method of dispute resolution — presentation of competing points of view to a neutral arbiter for decision according to rules of evidence, presumptions of innocence, burdens of proof, etc. While often effective for resolving disputes, the adversarial system has severe limitations when applied to fact-finding. What is often called an "inquisitorial" approach is usually better suited to this purpose.[3] In a particular case, both approaches might have to be employed separately. A "whistle blower," for example, might have a genuine personnel-related dispute with a manager or co-worker, but also have knowledge of serious misconduct that could affect the institution's scientific work. Resolving the personnel issues will not necessarily resolve the allegations of misconduct, which may be

---

[3]In an inquisitorial system the fact-finder plays an active role in gathering evidence. Additionally, instead of relying on adversarial parties to present competing versions of an issue, the inquisitorial fact-finder examines all sides of an issues on his own initiative. Potentially adversarial parties are given an opportunity to comment on and supplement the record, but are not primarily responsible for its creation, as they are in the adversarial system.

such that the scientific institution and in some cases the public have an interest overriding that of the whistle blower. Even in the extreme case in which the whistle blower is found to have made an allegation maliciously, it will still be necessary to demonstrate that the merits of the allegation have been addressed.

B. Integrating Human Behavior and Technical Aspects of an Investigation

While it is often necessary to separate fact-finding and dispute resolution issues, it is a common mistake to allow the human behavior and technical aspects of an investigation to proceed on separate tracks. If the alleged misconduct relates to a technical subject, it will be obvious to most investigators that some assistance from experts in that field is required, if only to understand the context in which the alleged misconduct took place. Less obvious is the need to integrate skills in reconstructing human behavior, such as through cross-examination of witnesses, with the technical issues that may have arisen from the allegation.

For example, if the allegation relates to an improper test procedure, technical assistance may be required to understand the procedure itself and how someone deviated from the correct practice. It may also be important, however, to establish why the mistake was made and, for this purpose, the technical analysis alone will usually not be adequate. The testimony of witnesses about their own conduct and that of others may be crucial to understanding the cause of the error, and these witnesses frequently have a strong interest in misrepresenting or obscuring the facts. Thus, skill in eliciting evidence under these conditions must be integrated with technical knowledge to establish a factual record sufficient to resolve the issue.

C. Standards of Conduct

An allegation of misconduct implies that there is a recognized standard of conduct that has been breached. As elementary as this sounds, institutional responses to allegations of misconduct often lose sight of the standards that should guide the inquiry. This can occur because the standards themselves are ambiguous or not widely known, or because, being human, the investigators allow their own subjective judgments to replace standards derived from a more authoritative source.

Standards are not always set forth in laws, regulations, and other authoritative legal sources. The customs and methods used by a scientific community in conducting peer review is a possible standard against which conduct in a particular case can be measured. Whatever its source, an explicit definition of the standard involved is important not only to ensure fairness, but to help determine whether dispute resolution or fact-finding is required and, in some cases, whether government intervention is warranted.

For example, one allegation that occurred in a technical environment was investigated internally, reinvestigated by a federal regulatory agency, publicized in the news media, and later recycled through litigation and petitions by activist groups. The allegation involved "harassment" and "discrimination," but the investigative reports and news accounts did not make clear what legal, administrative or other standards were being used to define concepts such as "harassment." It was strongly implied, however, that an employee had experienced harassment as a result of expressing safety concerns that would be of interest to the public and would justify action by a government agency. A closer examination of the evidence revealed that there was not even an allegation of this type of harassment, much less proof that it had occurred. Instead, the "harassment" and "discrimination" related to personal conflicts in an office environment, none of which in any way implicated safety systems or involved punishment of a whistle blower for expressing safety, quality, or other technical concerns. The failure to focus on the appropriate standard of conduct resulted in years of unnecessary controversy about a matter that should have been resolved without government intervention.

D. Response of Management and Other Systems to Allegations of Misconduct

Allegations of misconduct that arise in a scientific setting, like those in other institutions, can be true or false, exaggerated or understated, important or trivial, well-intentioned or rooted in jealousy, revenge, or other base motives — or a complex mixture of the above. Regardless of what the evidence demonstrates concerning the truth or falsity of a specific allegation of misconduct, the allegation itself nearly always poses a test for the various control mechanisms within an organization that are designed to prevent and correct errors, and to ensure the quality and integrity of the organization's product.

At one extreme, an allegation of misconduct might turn out to be both serious and true. Proof of such an allegation will invariably focus on the motives and actions of the guilty individuals, but the institution must also address other issues, such as the following:

- To what extent did the misconduct affect the scientific work of the organization? Will research have to be repeated, disclaimers issued, etc.?

- Did institutional practices contribute to the misconduct? For example, was the guilty individual subjected to undue pressure to produce results for fundraising, political, or other purposes? Was there an institutional "culture" that encouraged questionable research practices?

- Why did institutional mechanisms for detecting and correcting errors, such as peer review, not detect the misconduct in question before it became part of a whistle blower's allegation?

- How effective was the institution's overall reaction to the allegation, once it surfaced?

These and similar questions will arise even if the allegation of misconduct proves to be untrue, exaggerated, or trivial. An institution's mishandling of an untrue or trivial allegation can become a more important and damaging issue than the allegation itself. Indeed, the perception that an institution has covered up or not responded to an allegation will often lend more credibility to that allegation than it deserves. Conversely, demonstrating that the organization has analyzed the performance of its own internal control systems will help reassure government agencies that the institution is capable of resolving misconduct allegations with little or no government intervention.

E. Suggestions for Avoiding Common Errors in Internal Investigations

To avoid the errors discussed above, the following guidelines are suggested:

1. *Address the Merits of the Allegation*

Even if there is strong reason to believe that the person making an allegation of misconduct is acting in bad faith, the merits of the specific allegation must be addressed impartially and objectively. This may be difficult to do in the emotionally charged atmosphere that often results from charges of misconduct. Especially when the target of an allegation is highly respected, it is tempting to focus an investigation on the motives of the accuser, rather than on the specific allegation. Only by addressing the merits of the allegation, however, can the institution ensure that, for its own internal purposes as well as to retain the confidence of governmental or other oversight entities, it has adequately responded to information revealing possible flaws in its scientific work.

2. *Analyze the Allegation According to Fixed Standards of Conduct*

If, for example, the allegation is falsification of research data, by what standards was the research to have been conducted, and in what way did the researcher allegedly violate those standards? In addition to being important for the reasons discussed above, focusing on standards helps ensure that the inquiry will not become unduly influenced by subjective impressions concerning the motives and personalities of the accused and accusers.

3. *Determine Whether the Issue Must Be Resolved Factually*

As discussed above, some misconduct allegations require a rigorous fact-finding to resolve issues of importance to the public, or that have implications for the integrity of the institution. Other issues, however, may require an adversarial approach or may be amenable to mediation, negotiation, and other less formal dispute resolution methods.

4. *Use a Team Approach That Integrates Technical and Human Behavior Issues*

To ensure that technical and human behavior issues do not proceed on separate tracks, consider a multi-disciplinary team approach to conducting an internal investigation. Depending on the nature of the allegation, the team might include legal, human resources, and accounting, as well as technical personnel.

5. When Fact-Finding is Required, Ensure That the Investigation is Adequately Documented

Ideally, the document resulting from the investigation should satisfy a skeptical government agency or other oversight group that the institution has responded to the allegation in a way that makes further investigation by outsiders unnecessary. At a minimum, the report should make it clear to the outside reader:

- The nature of the allegation and the context in which it arose;

- The standards of conduct involved;

- The evidence relied upon by the institution to support its conclusions;

- The reasoning process leading from the evidence to the conclusions, including an explanation of how any conflicts in the evidence were resolved;

- An analysis of root causes of any misconduct or other flawed or abnormal conditions uncovered by the investigation;

- An analysis of line management responses to the allegation, as well as the responses of other entities within the institution (such as a peer review group), which are charged with evaluating charges of misconduct.

III. Regulating Scientific Conduct in the Government Research Organization

The federal government does not seek to control research practices in a way that could do harm to the integrity of the research process; on the contrary, ensuring that the integrity of that process is maintained is one of the primary objectives of government oversight. To carry out this objective, the government does seek to impose some controls on the behavior of those who work in federally-funded research institutions because there is a clear public interest to be served in doing so.

While the research community has an understandable concern that inadvertent harm will be done by government intervention, it is easy to forget that very few, if any, allegations of misconduct in science arise from federal government officials; they arise from within the scientific community itself. Government has an obligation to respond appropriately to allegations of misconduct to protect not only the integrity of the research process, but to protect the rights and freedoms of individual scientists.

A panel of the National Academy of Sciences has suggested that where there is no preexisting tradition of governmental regulation (such as there is in the case of laboratory safety, the treatment of human and animal subjects, or the use of toxic or hazardous substances), the federal role in regulating misconduct in science should be restricted to issues involving fabrication, falsification and plagiarism in proposing, performing or reporting research. Other questionable research practices and other forms of misconduct should either be subject to internal institutional controls or be regulated in accordance with existing statutes (e.g. those that deal with gross negligence, harassment, vandalism, etc.). One important problem that was alluded to in the NAS panel's report, but not directly addressed, is the fact that some questionable research practices (as defined by the NAS panel) could be used to cover up instances of misconduct in science. For this reason, if for no other, the government may find it very difficult to draw a hard and fast line between investigating misconduct in science and investigating questionable research practices.

Another important issue that was not directly addressed by the NAS panel concerns the role of the government in regulating research at government research institutions, such as those owned by the Department of Energy. The government has a special responsibility for the working conditions of scientists at these institutions. If the government owns an institution, regardless of whether the work conducted there is conducted under contract or by federal employees, the government has more of a responsibility to control the environment in which research is done than in institutions to which the government merely lends its support in the form of research grants.

It is our view that the government should adopt a total quality management approach to regulating scientific conduct at government research organizations, and therefore it should:

- clearly define what behavior is considered misconduct;

- develop and implement a program of continuing education for those whose behavior is thus proscribed, educating them in the definition of misconduct, the process for adjudicating allegations of misconduct, and the penalties that can be imposed for the various forms of misconduct;

- establish an open, predictable and fair process for handling allegations of misconduct, ensuring due process, objectivity and fairness to the accuser and the accused;

- recruit, select, train and manage a staff of investigators and hold them accountable for ensuring objectivity and fairness in every investigation;

- recruit, select, train and manage other staff to weigh the evidence and propose penalties appropriate to the facts in a given case;

- monitor and audit institutional programs aimed at achieving self-regulation of scientific conduct and recommend improvements in such programs based on the findings of those evaluations;

- measure, analyze and improve the performance of both the institution and the agency in preventing and responding to allegations of misconduct;

- establish nonmonetary rewards for good institutional programs of self-regulation and good agency programs for investigating and adjudicating allegations of misconduct; and

- ensure the overall process is continuously managed in accordance with a philosophy of total conformance to the requirements of prescribed behavior through quality management of the research process.

From our perspective, there is no good reason, other than our great difficulty in managing organizations effectively, why these institutional and governmental responsibilities cannot be discharged without imposing orthodoxy on the research community that would neither be in their best interests nor in the interests of the government.

Improving Mentoring

R. E. Fornes

A mentor is a person who provides guidance to another and, to some extent, initiates the person into the discipline. The guidance takes on various forms and degrees where the mentor may serve as counselor, teacher, coach or inspirer to the mentored. The mentor is usually older, of higher stature, more experienced and often at a higher skill or intellectual level than the apprentice.

The term mentor is found in Greek Mythology. Mentor was the faithful confidant, friend and trusted advisor of Odysseus in Homer's Odyssey. He was entrusted to teach and provide guidance to Odysseus' son Telemachus while Odysseus was on his long journey. Over the years, it has become accepted that mentoring is an important way to pass the knowledge obtained by the wise onto the inexperienced.

Just as in parenting, there is no book or set of formulas on mentoring that guarantees success. There is a perception that the "scientifically rich get richer"[1] by the selection process. Certainly, it is widely believed that much difficulty remains for women and minorities to break into the inner circles in areas traditionally dominated by white males, and that the white male mentors still show a lack of sensitivity to persons in these two groups. For example, earlier this month the National Research Council's (NRC) Committee on Women in Science and Engineering held its second annual conference, which was entitled "Women Scientists and Engineers Employed in Industry: Why So Few?" The importance of "mentoring and role models are crucial elements in bringing more women into science and engineering" was highlighted in the discussion.[2] In addition, the Association for Women in Science has just announced publication of *A Hand Up: Women Mentoring Women in Science*, in which the entire issue is devoted to mentoring.[3] At last week's AAAS meeting in Boston, a special session was devoted to "Mentoring and How it Impacts on Women in Science, Engineering and Mathematics."[4] In recent years significantly more financial assistance has been made available by the federal agencies, foundations and the private sector to address some of the inequities and shortcomings. However, as noted in some of the discussions at the AAAS and the NRC meetings, much work lies ahead.

Note that in the academic disciplines the nature of professional interactions and the manner in which mentoring is accomplished are strongly discipline dependent. A few years ago, I was co-author of a short paper published in the *American Journal of Physics* in which a simple analysis of the nature of collaboration as a function of discipline was done. We were interested in examining the extent of interaction, by some quasi-quantitative measure, between graduate students and their research advisors and others in their research environment. We made the assumption that, if we were to examine authorship, for example, then the greater the number of authors per paper, the greater the degree of collaboration. To obtain a quantitative assessment, we selected for comparison the following disciplines: modern language, philosophy, political science, business, psychology, education, botany, chemistry and physics. We contacted scholars in each field who referred us to journals of high quality in that area (we asked them to identify the journal considered the most prestigious). We selected one hundred full articles (to ensure statistical significance) for a given year from each journal starting with the last article and counting backwards. The results are as follows:

Field[6]	AVG # Authors/Paper
modern language	1.03
philosophy	1.03
political science	1.39
business	1.59
psychology	1.73
botany	1.80
education	2.11
chemistry	3.24
physics	4.87

We observed the large difference in the physical sciences — especially physics and the other fields. We investigated physics further by selecting six subfields in *The Physical Review* and counted the authors of thirty articles. The results were:

Physics Subfield[7]	AVG # Authors/Paper
nuclear	5.63
atoms and molecules	3.33
fluids and plasmas	3.23
condensed matter/structure	3.73
condensed matter/electronic properties	3.07
elementary particles*	11.2
(*theory	2.17)
(*experimental	28.03)

The results show the extreme case for experimental elementary particle physics which is obviously associated with costs and complexity of experiments where teamwork and careful coordination are generally required. We entitled the paper "Physics as a Team Sport."

The data clearly show the strong field dependence on the extent of collaboration. Therefore, any attempt to address issues in mentoring must take into account the field as well as the gender and ethnic origin of the partcipants.

References

1. Robert Kanigel *Apprentice to Genius* MacMillan Publishing Company, New York p. 238, 1986.

2. Mairin B. Brennan, "Women Scientists, Engineers Seek More Equitable Industrial Environment," *C&E News*, p. 13-16, Feb. 8, 1993.

3. *A Hand Up: Women Mentoring Women in Science*, Association for Women in Science, 1993.

4. *AAS 93 Science and Education for the Future* AAS Publication 93-03S, pp. 191-192, 1993.

5. J. D. Memory, J. F. Arnold, D. W. Stewart and R. E. Fornes, Physics as a Team Sport, *Am. J. Phys.* 53:270, 1985.

6. Data summarized from reference 4

7. Data summarized from reference 4

Some Considerations in Defining Misconduct in Science

Donald E. Buzzelli, Office of Inspector General, National Science Foundation

There can be several reasons for wanting a definition of misconduct in science, but I will concentrate on its use by government agencies when they process misconduct allegations. Within the federal government, both the National Science Foundation (NSF) and the Public Health Service (PHS) have issued regulations under which they deal with misconduct cases involving their grantees and grant applicants. These regulations contain a definition that states the range of actions that the agency will treat as misconduct in science.[1] Most universities have adopted their own definitions, many of them patterned after the NSF and PHS definitions. I will be speaking from the perspective of the Office of Inspector General at NSF, which receives and investigates all NSF misconduct cases. However, my remarks will be personal rather than official statements.

The purpose of this brief talk is to point out some of the major questions that must be answered if one tries to define misconduct in science. I also want to apply those questions to the current controversy about the definition. Much of this controversy is centered on a report by a panel convened by the National Academy of Sciences.[2] That report is part of the agenda for this session. The report is critical of the NSF and PHS definitions and recommends certain changes. According to one interpretation of the panel's recommendations, the definition should include falsification, fabrication, and plagiarism, and nothing else. I am not opposed to changes in general, but I think that in this case the recommended changes are not well grounded, largely because they do not address the fundamental questions.

Perhaps the most basic question is what makes any action misconduct in science. If one wants to promote a list like falsification, fabrication, and plagiarism as being the full range of misconduct in science, the first question is why those things are misconduct in science at all. What character or property do those actions have, if any, that makes them misconduct in science? If one can answer that question and can find such a property, one can then ask whether

---

[1]The NSF definition is: " 'Misconduct' means (1) fabrication, falsification, plagiarism, or other serious deviation from accepted practices in proposing, carrying out, or reporting results from activities funded by NSF or (2) retaliation of any kind against a person who reported or provided information about suspected or alleged misconduct and who has not acted in bad faith." The PHS definition is: " 'Misconduct' or 'Misconduct in Science' means fabrication, falsification, plagiarism, or other practices that seriously deviate from those that are commonly accepted by the scientific community for proposing, conducting, or reporting research. It does not include honest error or honest differences in interpretations or judgments of data."

[2]National Academy of Sciences, Panel on Scientific Responsibility and the Conduct of Research, Committee on Science, Engineering, and Public Policy, *Responsible Science: Ensuring the Integrity of the Research Process* (National Academy Press, Washington, DC, 1992), vol. 1, esp. pp. 24-30.

there are other actions that also have that property and should be included. In this way, one can judge whether one's definition is complete. The Academy panel produced one such criterion, namely that misconduct in science is any behavior that seriously damages the integrity of the research process. Unfortunately, this insight was not developed or fully applied. If that were done, I believe it would turn up other actions in addition to falsification, fabrication, and plagiarism.

A second question is what is the relation between a formal definition of misconduct in science and the standards of conduct held by scientific community. Does the definition express all, or any, of the scientific community's standards for the ethical conduct of research? Most people would agree that it expresses those standards in some way, but in developing a definition it is necessary to go farther and take a position on whether one wants to bring in all the major requirements that the community has for the ethical conduct of research. I will say more about this later. Again, I think the panel's discussion of this point was incomplete. I also do not believe that falsification, fabrication, and plagiarism exhaust the community's standards for ethical research.

A third question has to do with the nature of the government's responsibility with regard to misconduct in science. Just how far does that responsibility extend? Federal granting agencies may notice many undesirable behaviors on the part of their grantees and grant applicants. Which behaviors do the Congress and the public expect the agencies to take action against as custodians of public funds? It is easy to ask an agency to limit its interest to falsification, fabrication, and plagiarism, or whatever, but it is not so clear that the agency is even free to do that. Again, I did not see that discussion in the panel's report, nor do I think that the agencies are expected to deal only with fabrication, falsification, and plagiarism.

After this, many more specific questions come up, but I will mention only one. Much is made in the report of the vagueness or ambiguity that the present NSF and PHS definitions are supposed to have. Under these definitions, scientists supposedly don't know what the government is likely to punish them for, and there is nothing to prevent the government from creating bizarre and abusive misconduct in science cases based on something a scientist does that really is perfectly innocent. To remove this vagueness the panel's report recommended that the phrase "other serious deviation from accepted practices" be removed from the NSF and PHS definitions.

This issue has many parts. I don't believe that the NSF and PHS definitions are excessively vague, but they clearly leave room for interpretation. This applies to all the terms in the definition, not only to the "other serious deviation" phrase. For example, what are plagiarism and falsification exactly? If the phrase "other serious deviation" were removed, I think the probable result would be to increase the vagueness of the terms falsification, fabrication, and plagiarism. Certain kinds of misconduct that now fall under "other serious deviation" would be forced under these other headings, so that these headings would lose whatever definite meaning they have.

Moreover, there will never be a definition that creates no problems of interpretation. The goal of creating a perfectly precise definition is unreachable. The only question is who will interpret the definition, and under what procedures. The NSF definition refers to the community of scientists as the ultimate interpreter of the definition. If this reference to the community is dropped, and the terms of the definition are left to stand on their own, the likely result is that definitions will require further definitions, and so on, and the whole problem of interpretation will fall to legions of lawyers.

The procedures for interpreting the definition enter in here. It is possible to imagine all kinds of abuse occurring under agency definitions if one does not consider the context in which they are used. In fact, NSF procedures involve a division of responsibility between the Office of Inspector General and the Office of the Director such that abusive cases are virtually certain to be rejected in one office or the other. The procedures an agency employs are much more effective in preventing abusive cases than any proposed change in the definition would be.

Lastly, I maintain that the so-called vagueness or ambiguity of the NSF definition is needed in order to deal with cases that one could not have thought of in advance. One of the basic questions in designing a definition is whether it should be closed-ended or open-ended. By this I mean whether it should consist of a finite list of types of misconduct only, or whether it should have some provision like "other serious deviation from accepted practices" that anticipates an unspecified range of types of misconduct. I believe it is not possible to write a list of all the kinds of misconduct in science that one may have to deal with in the future. Only an open-ended definition can be flexible enough to anticipate all the cases an agency may have to deal with. The panel report vigorously complained about the specific wording of the "other serious deviation" passage, but it is less clear whether it was taking a stand against all open-ended definitions.

The proposal to drop "other serious deviation" from the NSF definition and retain only falsification, fabrication, and plagiarism seems to be based on the assumption that the definition is a list of four independent items. However, the NSF definition is basically not a list at all. Rather, the definition simply says that misconduct in science is deviation from accepted practices. With this definition the agency is expressing its intention to treat as misconduct in science anything that seriously deviates from practices accepted by the scientific community, in connection with research and education in science and engineering. Falsification, fabrication, and plagiarism are given as examples of what such serious deviations are, and they limit the interpretation of that phrase. If anyone does not know what a serious deviation from accepted practices is, or has worries about what it means, the definition says it is the sort of thing illustrated by falsification, fabrication, and plagiarism. The proposal to remove the "other serious deviation" phrase is a proposal to retain the examples and remove the thing they are examples of.

Any proposal to write a definition of misconduct in science has to consider the unique kind of offense that misconduct in science is. Our office has dealt with people who admitted to felonies but still resisted being charged with misconduct in science. Any scientist would understand that. A scientist who commits a crime is a bad person, but a scientist who commits misconduct in science is a bad scientist. In some ways, being a bad scientist is worse. It means that the scientist has disgraced the profession to which he or she may have devoted a lifetime.

Therefore, misconduct in science should not be thought of as something the government made up and put into a regulation. When the government accuses someone of misconduct in science, that person is not simply being accused of violating a regulation. The accusation really means that the person is a bad scientist. That is why this subject is so emotional. Misconduct in science involves a stigma in the scientific community because the individual allegedly has violated the standards of the community and has merited rejection by the community. However, the standards for being a good or bad scientist do not come from the government. They reside in the scientific community prior to any government rule-making.

These considerations seriously limit any definition of misconduct in science that a government agency may issue. The agency cannot put out a regulation telling the community what the community's own standards are. That is why the definition does not take the form of a list of all the things a scientist should not do. Moreover, any possible definition is limited to those things that the scientific community would recognize as making someone a bad scientist. A definition that does not meet that requirement should not be labeled misconduct in science.

Furthermore, the definition should cover all the actions that violate the standards of the community and make someone a bad scientist. Misconduct in science is whatever violates those standards. Absent any argument to the contrary, the government is responsible for dealing with the full range of misconduct in science. Hence, there is also a requirement of completeness that the definition must meet.

The NSF definition meets these requirements in a simple way, by referring to the practices accepted by the community as its standard for what is included or excluded. There may be other ways of meeting these requirements, but the current proposals I have heard for replacing the NSF definition do not meet the requirements. For example, the list falsification, fabrication, and plagiarism does not meet the requirement of completeness.

Let me conclude by summarizing the points I have been trying to make. Negatively, I suggested that "falsification, fabrication, and plagiarism" is not a complete list of any kind. Positively, I suggested that there are basic questions that have to be answered before one can produce a sound definition. These have to do with what misconduct in science basically is, and how it is related to the standards of conduct of the scientific community and to the responsibilities of federal funding agencies. I suggested that there are criteria for distinguishing misconduct in science from other offenses that a scientist may commit. One very basic criterion is that misconduct in science comprises all those actions, and only those actions, that the scientific community would regard as making someone a bad scientist. I argued that the definition should be open-ended in order to anticipate all the important cases that may come up. Finally, the scientific community has a right to be assured that federal agencies are not going to create misconduct in science cases based on some scientist's perfectly innocent behavior. I suggested that this assurance should be sought in the procedures that agencies use, more than in definitions.

Science and Media

Deborah Blum

The area that interests me most at the moment — probably because I'm living with it — is that of how scientists respond to requests for information on controversial or complex subjects.

You can make the case, of course, that that covers everything in science.

But, I've spent the last couple of years researching and writing about animal research, an obviously controversial topic, and I want to use that experience as a kind of model for some critical issues.

I am coming to think that one of the major problems for science writers is that we tend to present science as a monolithic thing — the research community, as if it was a bland, blank fortress. And, of course, it isn't. Let me give you two examples out of research from a book I am writing on primate research. I was at one primate facility where the chief vet, as we approached the buildings, said to me "You're going to hate this place. Every day I wish I could blow it up myself." It was because he genuinely cared for the animals so much. At another, they were doing leprosy work on sooty mangabeys. And one of the mangabeys was extraordinarily affectionate; the technicians could take her out of her cage and she would wander up and down the halls, hugging people. And the director of that facility was extremely irritated that she hadn't been infected yet with leprosy. He saw no value in her outside of research.

We have to learn, I think, to present science in all its diversity. And to do that well, will take cooperation and willingness from both journalists and scientists.

So, my first point is fairly pragmatic. Because I write about animal research, a field muddled in controversy and politics, many people don't want to talk at all. The head of the Michigan Society for Medical Research told me recently that they got a request for help on an animal research story. She called 37 researchers and they all refused to do the interview.

When I was writing about primate research for the *Sacramento Bee*, I wanted to visit a major pharmaceutical company that is the largest importer of monkeys in the state. I knew that because I'd spent almost a year filing public record act requests. When I called them, they dickered about it for two months. At one point, they were offering anonymous phone calls in the middle of the night. And finally they said no anyway.

I lost my temper, and then I called everyone I knew to put pressure on. And they reversed themselves. But to tell you how reporters work, by the time they called, I was sorry they did. I had a working outline of a wonderful story about how paranoid animal researchers were. Instead, I had to write a story about how open they are. Not exactly, of course, but it had a different edge. And I hope you can see that the researchers served themselves much better by being open.

I know many scientists who really dislike operating in the public arena. But I believe that scientists should be held accountable to the public. Not just because the public often pays for the work. But because science can so impact people's everyday lives. And yet, there are some interesting problems involved with that. For instance, you would be astonished at how much research work can be accessed through public records — which is exactly what the animal rights people do. I know scientists now who are censoring what they write in their internal lab records, out of fear that they will become public. And I think we should be concerned when researchers become too self-conscious about their work. In a free press society, the balancing of public interest and privacy is always a difficult one.

But despite that, we are fortunate to have a free press, uncontrolled by government, uncensored by political interests. At least, that is our aim. Or to put it another way, I think you're stuck with us. Short of going door to door, we are the best way to tell your story. The onus on you is several fold; how can you use us to tell your story well and accurately, how do you develop a relationship of trust.

And I want also to make the point that in some ways science writers, like myself and David Perlman, (science editor of the *San Francisco Chronicle*, also on the panel) may be a little more attuned to the concerns of scientists because we specialize in science journalism. But as the public gets more interested in science, reporting on it is becoming an open field. I was on a panel last summer of the Investigative Reporters and Editors conference and there was widespread interest in investigating science. In fact, some of the splashiest science stories have not been done by science writers. Consider John Crewdson of the *Chicago Tribune* and his investigations of Robert Gallo, or the very fine piece on the problems of the Hubble Telescope, by the *Hartford Courant*, which won a Pulitzer Prize. That work was done by general assignment reporters.

Like science, journalism is diverse, and we must learn to understand each other on several levels. I think, I hope we are all becoming more sophisticated in that understanding. I hope you will recognize that each encounter with the media does, however incrementally, deepen that sophistication. Finally, a word about newspapers. Yes, we are built for simplification. And that's getting worse, not better. But, if nothing else, perhaps we can learn to tell our simple stories better.

There are two things I use as indicators of improvements in telling the story of science. One is that I'd like, in my own newsroom, a little more competition for my job. I think I have a great job. I'd like to see other reporters covet it more, be a little less fearful of covering science. The other point has to do with basic science. We still are far from being able to present that well, and by well, I mean with a sense of wonder. We do pretty well with applied work, but when we are really able to get people jazzed about basic curiosity, then we have come a long way.

Teaching Ethics in the Sciences: Why, How and What[1]

Stephanie J. Bird, Ph.D.

There has been growing awareness of the importance of addressing the ethical issues and values associated with the professions. Biomedical ethics has become an accepted, even expected component of medical education and business schools also have established courses and programs in ethics. Many engineering and computing societies and the American Chemical Society have codes of ethics. However among professional societies in science and technology, these codes of ethics are the exceptions rather than the rule.

Over the last decade concern about ethics in the scientific professions has arisen, in part, as a result of instances of egregious misconduct that have threatened both the reputation and the fabric of the professions. In addition, highly publicized problematic practices have heightened awareness of the potential for misunderstanding, confusion and conflicts of interest and values.

Publicity associated with actual or perceived instances of misconduct in science has shaken public and governmental confidence in the efficacy of informal methods of maintaining high professional standards in scientific research. Indeed, increased Congressional concern led to the establishment of the Office of Scientific Integrity of the National Institutes of Health and the formation of a panel of the National Academy of Sciences to examine issues of scientific integrity.

As a reflection of these concerns, in 1989, the Alcohol, Drug Abuse, and Mental Health Administration (ADAMHA) and the National Institutes of Health (NIH) instituted a requirement that training grant applicants for National Research Service Awards (which provide for the training of graduate students) include "a program in the principles of scientific integrity [as] an integral part of the proposed research training effort."

In addition, there is an often unarticulated expectation on the part of students, the public and professional colleagues that a complete education in the sciences and engineering should include courses or programs that address matters of professional ethics. There is increasing awareness of the responsibility of professionals to train students and younger colleagues not just in concepts and techniques, but also in the standards of the profession. Within the scientific community, there is concern that the time-honored techniques for transmitting professional standards are inadequate in the face of rapidly increasing numbers

---

[1]Some material presented here was included in the conference/workshop "Teaching Ethics in Science and Engineering: Why, What and How" held February 10-11, 1993 in conjunction with the annual meeting of the American Association for the Advancement of Science in Boston, Massachusetts. I also want to acknowledge and thank my colleague Caroline Whitbeck for sharing her insight into many of these issues.

of scientists which exceed the availability of research funding, the fast pace of many areas of research, and the expanding potential for linking basic research to profit-making applications. Preliminary reports indicate that graduate students believe that the best way to learn professional values and standards is through their mentor(s) (J.P. Swazey, personal communication, August, 1991).[2]

Science professionals also recognize that the behavior of their colleagues is a reflection on them. They have much to gain by counteracting widespread public doubts and misperceptions. Thus there are internal and external pressures on the scientific community to develop educational programs that address these issues.

Why teach ethics in science?

There are a number of strategies that can be used in raising and addressing ethical issues in the life sciences. Whichever is employed, a critical component for success is the recruitment of colleagues. This is important because the extent to which the faculty are involved is likely to have a significant impact on how seriously students, postdocs, and other colleagues regard the issues. Students learn from their mentors and those they wish to emulate.

It is crucial that scientists recognize that addressing ethical issues and concerns is an important component of education since students look to the faculty to identify the important elements of the profession. Since predoctoral students and, to a lesser extent, postdoctoral fellows regard advisors and lab heads as role models, involvement of science faculty in the development and implementation of whatever programs are proposed is essential for credibility with students and junior investigators. Educational programs need to be supplemented and complemented with informal approaches that emphasize the continuing relevance of high ethical standards in every day research. Senior investigators are key in creating a sense of professional responsibility and an atmosphere of openness where questioning, discussion and healthy skepticism can flourish.

Discussion of professional values and standards among colleagues has the added benefit of promoting and facilitating identification of standards, values and conventions. Such discussion helps to clarify accepted differences in values and the underlying assumptions upon which they are based, as well as unacceptable practices. This has not usually been part of professional education, but rather has generally been considered outside the realm of the discipline. Thus professionals rarely reflect on values and expectations, nor professional standards and their implications.

---

[2]Judith P. Swazey, President of the Acadia Institute for the Study of Medicine, Science and Society, and Karen S. Louis, Associate Professor of Educational Policy and Administration at the University of Minnesota, are currently engaged in a National Science Foundation funded study to examine professional values and ethical issues in graduate education. Their findings are more fully described in Dr. Swazey's background paper prepared for this session "Teaching Ethics: Needs, Opportunities, and Obstacles."

Methods of teaching ethics

Different approaches for presenting ethical issues include formal courses, departmental or intra-institutional seminars, and integrating discussion and analysis of ethical issues into established core courses. Each approach has advantages and disadvantages.

A course emphasizes the importance of the topic with a grade and can have the advantage of covering subjects more thoroughly. A course also can make it possible to cover more subjects systematically than a piecemeal, occasional lecture or seminar. However, a course has the disadvantage of potentially being marginalized. It may seem to have the effect of absolving faculty members, and those not involved, from the need to address ethical issues in their own courses or other work. On the other hand, its value can be enhanced depending on which and how many faculty participate in the course (e.g., the chairman of the department), its role in the curriculum (e.g., it is a required course or a limited elective highly recommended by faculty advisors), and the extent to which its messages and topics are reflected and reinforced in other courses.

Departmental seminars stress the importance of these issues for, and to, the whole community and provide an opportunity for many more faculty to be involved, to share their views, and to clarify the expectations of professional colleagues and the standards of the profession. This can be informative for postdocs and other faculty, as well as students. Departmental seminars also allow a broader range of individuals to participate in the discussion, e.g., research, technical, clerical and other support staff. This is particularly valuable since support staff are often left out of the discussion, and their ethical concerns are generally not taken into consideration.

However, if most faculty do not attend or participate in departmental seminars that focus on ethical issues, they convey the message that ethical issues are not really important. In addition, it is likely that fewer topics will be covered in a seminar format. It also makes it possible to relegate ethical issues to the margin of science by making it an occasional topic of interest.

Integrating ethical concerns into courses, especially core courses, underlines the fact that ethical issues are inherent rather than tangential to the discipline, and to the profession. However, if the material is not a part of the course grade, it can send the opposite message. In addition, some important topics may not be covered because they do not fall obviously within the purview of a particular subject, e.g., authorship, safety, discrimination in the workplace.

Ideally, a combination of approaches would provide the greatest likelihood of achieving the goal of wider awareness and sensitivity to ethical issues and implications of our work. Programs of all types should be supplemented with the example of the faculty. However, it should be noted that, as indicated by Judith Swazey in her background paper, although over 60 percent of faculty believe that the most effective, indeed the only effective approach to teaching ethics is through interactions with faculty in research and/or informal discussion when issues or problems occur, graduate students have relatively little contact with even "particularly supportive" faculty (i.e., those faculty members that students identify as expressing continuing interest in their progress, providing letters of recommendation, or providing financial support) and receive little input regarding good research practices, teaching techniques, and the development of professional relationships.

In raising and discussing ethical issues, case studies or scenarios have been found to be an especially useful teaching tool. They catalyze discussion and highlight the ways in which ethical issues are interwoven into science, often arising out of confusion and misunderstanding about conventions, or responsibilities, or differing expectations, values or needs. Differences and misunderstandings are exacerbated by lack of communication and clarification.

Case studies and scenarios can also serve to emphasize two very important points. First, individuals need to be trained in assessing the source and focus of the problem and in designing a course of action that solves the problem for them. Rarely is either process obvious or easy given the complexity and interrelatedness of both science and human relationships. Second, in developing a solution to an ethical problem, there is likely to be more than one acceptable or even "good" solution. Invariably, there is at least one "bad" solution. However, what constitutes an acceptable or preferable solution is likely to depend on the perspective of the different individuals involved.

Introduction of ethical theory in discussing ethical issues has the advantage of introducing a body of literature which often helps to clarify the issues, and of providing a common base of understanding and language with which to examine the topic. It also helps to prevent the discussion of cases from becoming a litany of "old war stories" that can imply that there is no organized way of thinking through ethical problems. However, too much emphasis on ethical theory can make the subject seem abstract, remote and inaccessible.

Topics in professional ethics in the sciences

The life sciences, and the work place and professional requirements of life scientists, provide a vast array of issues, concerns and implications. Topics specific to the work environment range from issues of safety to the relationship between those who work together, which can provide a setting in which sexual harassment or the expression of prejudice can occur. Inherent to the nature of scientific research are such topics as research design, data selection, authorship, peer review, the potential for conflict of interest, the impact of funding on the direction of research, and, in the life sciences, issues associated with the use of animals and human subjects in research. The larger implications of life science research include its appropriate role in the development of public policy regarding such diverse concerns as the global environment, health care, behavior modification and risk assessment.

Conveying the standards and expectations of the discipline is the responsibility of the entire scientific community including professional societies. As part of our responsibilities both within and beyond the scientific community, it is essential that students and science professionals at all levels be prepared to recognize and address the ethical issues that are inherent to nearly every aspect of science.

Teaching Ethics: Needs, Opportunities, and Obstacles

Findings from the Acadia Institute Project on Professional Values and Ethical Issues in the Graduate Education of Scientists and Engineers[1]

Judith P. Swazey, Ph.D.

A brief overview of the graduate education project, including its four data sets, is appended to this paper. For the Sigma Xi Forum, we have summarized some of the marginal analyses of data collected from graduate school deans, and from faculty and doctoral students in chemistry, civil engineering, microbiology, and sociology, about a number of matters related to "teaching ethics." We hope that these findings will help to inform efforts to incorporate systematic and sustained attention to professional and research ethics in various educational contexts.

Becoming Teachers and Researchers. Since doctoral students in major research universities are the principal sources of tomorrow's teachers and researchers in those institutions, we are interested in the types of value-laden messages students are receiving about the importance of teaching and research in their programs. This question also is relevant to understanding the views of faculty about the importance of devoting time to teaching ethically relevant materials either in courses or as part of the "everyday life" of research training.

1. There are striking differences in students' views about how carefully faculty supervise their teaching assistants (TAs) and research assistants (RAs). A substantial majority of respondents in the student survey (60 percent) *do not* think that TAs in their program are carefully supervised. Conversely, over 70 percent think that RAs are carefully supervised. With respect to disciplinary differences, civil engineering students feel most strongly that TAs and RAs are carefully supervised. Sociology students feel most strongly that TAs, and chemistry students that RAs, *are not* carefully supervised in their departments.

[1]The project is co-sponsored by the AAAS Committee on Scientific Freedom and Responsibility, The Council of Graduate Schools, and Sigma Xi, and supported by Grant No. 8913159 from the National Science Foundation. The following NSF components have provided funding to the Ethics and Values Studies Program for support of the project: the Directorate for Behavioral, Social, and Economic Sciences, the Directorate for Biological Sciences, the Directorate for Engineering, the Directorate for Mathematical and Physical Sciences, and the Office of the Inspector General. Any opinions, findings, conclusions, or recommendations are those of the author and do not necessarily reflect the views of the National Science Foundation.

2. Students also were asked whether, in their opinion, most faculty in their department "really care about their teaching." Slightly over half (56 percent) said "yes," and 44 percent "no." Male and foreign students were significantly more positive about faculty as teachers than women and U.S. citizens: only 45 percent of women and 48 percent of U.S. citizen students judge that most of their faculty really care about their teaching, compared to 59 percent of the male and 63 percent of the non-U.S. citizen respondents. By discipline, the percentages responding "yes" were: chemistry, 47; civil engineering, 70; microbiology, 55; sociology, 45.

   Sources of Graduate Students' Professional Values and "Ethical Preparedness." All of us have sets of values and ethical standards that are shaped by many sources and influences, from childhood through adulthood. Students were asked how important each of 10 different sources has been in shaping their professional values and preparing them to deal with ethical issues in their field. Not surprisingly, their responses show that their professional values and ethical preparedness have been shaped by many influences prior to and outside of the graduate education context as well as by graduate training itself.

1. Combining the percentages of "very" and "somewhat important" responses, the students' rank ordering of the 10 sources was as follows:

   1. Supportive faculty member(s)

   2. Other graduate students

   3. Family

   4. Friends/colleagues not in graduate school

   5. Faculty in undergrad program

   6. Other graduate faculty

   7. Religious beliefs

   8. Discussions in other courses, labs, seminars .

   9. Professional organizations in field

   10. Courses dealing with ethical issues

2. The low rankings given to discussions of professional values and ethical issues in contexts such as other courses, labs, and seminars, the role of professional organizations, and courses dealing with ethical issues are not "good news" for those engaged in efforts to stimulate ethics and values teaching. Other portions of our survey findings and our interviews, however, support the view that these sources are not unimportant per se, but rather that students have had relatively little exposure to them. With respect to the role of professional organizations, for example, few faculty and fewer students whom we interviewed knew whether their major professional organization had a code of ethics. Many interviewees said, "I assume it does, but I'm not familiar with it and wouldn't know where to find it." In

the faculty survey, 32 percent of respondents stated that they know their primary professional association has a code of ethics but they were not familiar with its contents; 16 percent did not know whether such a code exists (ranging from a low of 6 percent of sociology faculty to a high of 32 percent of microbiology faculty). In the interviews we also encountered very few faculty who said they have ever incorporated ethics and values topics into their courses or other teaching, apart from episodic discussion of current "hot topics" such as cold fusion or the Baltimore case, and correspondingly few students who said they had been exposed to ethics and values discussions or more systematic teaching in their undergraduate or graduate work. Faculty views about the most effective ways for graduate students to learn about the professional values and ethical standards and issues in their field are reported in the next section. These data quantitatively extend the interview findings, and mesh with and help to explain the student rankings of the most important sources of their professional values and "ethical preparedness."

3. The top ranking given by graduate students to "particularly supportive" faculty merits a brief explanation and commentary. In the survey, we asked students whether "there is at least one faculty member (including your advisor, if appropriate) in your department who is particularly supportive of you and your work." We did not define or even use the words "advisor" or "mentor" because the terms and concepts have so many different interpretations. The great majority of respondents (89 percent) said they do have one or more faculty members who play particularly supportive roles. Those students then were asked to respond to a list of 14 advisor- or mentor-like things a faculty member may do for a graduate student, assessing how much help their most supportive faculty gave them with each item ("a lot," "some," "none").

Considering the possible and often ideally depicted roles of research and dissertation advisors and mentors in graduate education, what is striking about the responses to this question is the low percentage of students who say they get "a lot" of help on most of the 14 items; this holds when the responses are tabulated by discipline, gender, citizenship and race. The most frequent types of activities or behaviors by "particularly supportive" faculty members, according to these initial analyses, are (1) to express continuing interest in a student's progress (the only item ranked "a lot" by over 50 percent of all respondent groups), (2) to write letters of recommendation, and (3) (except for sociology) to help students get financial support. Substantially smaller percentages of students reported that they get a lot of help on aspects of their doctoral education such as: receiving helpful criticism on a regular basis, learning the details of good research practice, advice about teaching, developing professional relationships with others in their field, and learning the "art of survival" in their field. These findings, as does other research, underscore three important points about advisors and mentors that are relevant, among other things, to assumptions about how professional ethics and values should be or are being transmitted to graduate students. First, it is fallacious to equate a mentor with an advisor or other person directly responsible for a student's research training and, second, therefore to assume that all graduate trainees have mentors. Third, as Baird points out, "although the *ideal* model of graduate education includes a great deal of student-faculty interaction," our study and other research show that there is little interaction in many areas that are important components of doctoral training and professional socialization, even with faculty whom students consider to be especially supportive of them and their work.

Faculty Views about the Most Effective Ways for Graduate Students to Learn about Professional Values and Ethical Standards and Issues in their Field

1. Faculty were asked to rate the effectiveness of 7 ways that students can learn about professional values and ethical standards and issues in their field, on a 4-point scale of "very," "somewhat," "not very," and "not at all effective." Their responses document an overwhelming belief that students most effectively learn these aspects of their professional work in the context of their research training, by example or role modeling, and by informal discussions of ethical problems "when they occur," rather than by more structured methods. By overall percentage responses for very effective, faculty rate the 7 ways of "teaching ethics" in the following rank order:

    1. Interaction with faculty in research work [65 percent very effective]

    2. Informal discussion of ethical problems when they occur [61 percent]

    3. Discussion of ethics and values in regular coursework [19 percent]

    4. Brown-bag session or colloquium [18 percent]

    5. Special courses devoted to these topics [14 percent]

    6. Department and university policies for teaching and research [12 percent]

    7. Codes of ethics and professional standards provided by professional organizations [7 percent]

2. Discipline-specific beliefs and attitudes about the effectiveness of various ways of "teaching ethics" need to be taken into account in developing strategies and efforts to more explicitly and systematically incorporate ethics and values into graduate training. Looking at our faculty survey data, for example, the highest percentages by discipline rating each of the seven items as "very effective" and "not very or not at all effective" were as follows:

	Very Effective	Not Very/ Not At All Effective
Discussion of ethics and values in regular coursework	Sociology [30%]	Chemistry [45%]
Special courses devoted to these topics	Sociology [16%]	Chemistry [65%]
Interaction with faculty in research work	Chemistry [75%]	CivEng [9%]
Informal discussion of ethical problems when they occur	Chemistry [68%]	CivEng [6%]
Brown-bag sessions/colloquium	Sociology [21%]	Chemistry [40%]
Professional organization codes of ethics and professional standards	Microbiology [11%]	All [56-54%]
Dept. and university policies for teaching and research	Sociology [14%]	CivEng & Micro [46 & 45%]

Faculty Assessments of their Students' "Ethical Awareness"

One of the reasons that many faculty are not actively or explicitly engaged in "teaching ethics" as part of their graduate student training activities may be because they believe that their students already know about the ethical standards and issues in their field. Our faculty survey asked respondents "what proportion of doctoral students in your department exhibit an awareness of ethical standards and issues in their discipline?" Overall, 27 percent of the faculty believe that all or almost all of their students exhibit such awareness, 47 percent that a majority of students do, and 32 percent that a minority, very few, or none do. By discipline, microbiology has the highest percentage of faculty (77 percent) who believe that most of their students are ethically aware of standards and issues in their field. Conversely, civil engineering has the lowest percentage (55 percent) of faculty who believe that most of their students exhibit such awareness. The bases on which faculty make such determinations is one of many topics that could be fruitfully explored by groups seeking to foster teaching activities concerning research ethics.

Can Values and Ethical Standards be Taught and Changed?
Faculty Perspectives

Our data also suggest two other reasons why faculty may be ambivalent about the importance of explicit ethics teaching, or believe that such teaching can be done best through the informal or latent learning that occurs in research training relationships. First, 40 percent of faculty respondents "strongly agreed" or "agreed" with the statement that "by the time students enter graduate school, their values and ethical standards are so firmly established that they are difficult to change." Substantially more civil engineering faculty than faculty in the other three disciplines hold this view (percent strongly agreeing or agreeing: civil engineering, 52; chemistry, 43; microbiology, 43; sociology, 24). Second, perhaps because they are unfamiliar with the substantive content of ethics and values studies in various professional fields, many faculty may not realize that there is "anything special" to be taught. The majority of faculty respondents in our survey (59 percent) strongly agreed or agreed with the statement that "it is hard to make a distinction between *professional* values and ethical standards and *personal* values and ethical standards." Thirty-eight percent disagreed/strongly disagreed with this statement, while 3 percent checked "don't know." By discipline, faculty positions on this matter were similar to their positions on whether a graduate student's values and ethical standards can be changed: civil engineering had the highest percentage of faculty (68 percent) strongly agreeing/agreeing that it is hard to distinguish between professional and personal values and ethical standards, compared to 47 percent of sociology faculty and 62 percent of chemistry and 61 percent of microbiology faculty.

Ethical Preparedness Training: High Importance, Low Activity Levels

1. Deans, faculty, and graduate students believe that what we call "ethical pre-paredness training" — preparing students to recognize and deal with ethical issues they may encounter in their field — *should be* an important function of their universities (deans) and departments (faculty and students). Overall, 99 percent of the deans, 88 percent of the faculty, and 82 percent of the student respondents said their institution/department should take a very active to somewhat active role in such training.

2. All three groups, however, report a substantial difference between "should" and "does" for ethical preparedness training. Overall, 49 percent of the deans judged that their institution is doing a *very or quite effective* job in this educational realm, while 51 percent felt that a not very or not at all effective job was being done. Faculty and graduate students, in turn, were asked how active a role their department *actually takes.* Only 3 percent of the students judge that their department takes a very active role and 19 per-cent an active role, while 25 percent report it is "not at all active." Faculty give their departments somewhat higher marks: 4 percent stated their department is very active, 37 percent that it is somewhat active, and 13 percent that it is not at all active in ethical preparedness training.

3. There are some substantial *disciplinary differences* in our findings, which bear on how receptive faculty in various fields in the sciences, social sciences, and engineering have been or will be about incorporating ethics into their teaching/training activities. Deans, for example, felt that engineer-ing disciplines in their institutions were doing the least effective and the social/behavioral sciences the most effective job of ethical preparedness training (68 percent rated engineering "not very or not at all effective,"

compared to 33 percent for social/behavioral sciences, 42 percent for life sciences, and 62 percent for chemical/physical sciences).

By discipline, sociology faculty responding to the survey feel most strongly (54 percent) and chemistry faculty least strongly (26 percent) that their department *should take* a very active role in ethical preparedness training; a very active role is endorsed by 37 percent of civil engineering and 44 percent of microbiology faculty. Conversely, 18 percent of chemistry faculty and 16 percent of civil engineering faculty believe their departments *should not* have a very active role or any role at all in this sphere of training, compared to 10 percent of microbiology and 7 percent of sociology faculty.

Ethical Aspects of Faculty Roles and Responsibilities

Faculty perspectives and activities related to "teaching ethics" — broadly defined as consciously or explicitly transmitting the professional values and ethical standards of one's discipline to students — are components of the ethical climate of a department or research group. Another component, which would seem to have a bearing on the importance faculty attach to ethical dimensions of their students' training, involves the extent to which faculty members believe they have and actually exercise professional responsibility for the professionally related ethical conduct of their trainees. Respondents to the faculty survey were asked to indicate (1) the extent to which they believe that faculty in their academic/research community should exercise a "collective responsibility for the professional-ethical conduct" of their graduate students, and (2) the extent to which faculty in their department *actually* exercise this responsibility. [Faculty also were asked these questions with respect to their departmental colleagues.] Almost all faculty respondents believe that to a great extent (74 percent) or some extent (25 percent), they do have such a collective responsibility. When faculty were asked to assess the extent to which they and their departmental colleagues actually manifest a collective responsibility, however, they reported striking differences between their professed values and actual practices. Only 27 percent judge that faculty in their department do exercise a great deal of collective responsibility for their students' professional-ethical conduct, and 61 percent believe that role is exercised "to some extent."

In terms of disciplinary differences, chemistry and microbiology faculty (79 percent) feel more strongly than civil engineers and sociologists (69 percent) that they have a very great collective responsibility for their students' conduct. There also are substantial differences in the extent to which faculty believe members of their departments actually exercise such a responsibility to a great extent: chemists believe they manifest the greatest amount (38 percent) and sociologists the least amount (18 percent) of collective responsibility.

Commentary

We are conducting various associational analyses to examine relationships between the findings reported in this paper. In terms of faculty data, for example, these analyses will include the relationships between faculty views about the most effective ways for students to learn about professional values and ethical standards and issues and about the importance of ethical preparedness training, the extent to which they believe that their graduate students already are aware of ethical standards and issues, and their views about whether students' values and ethical standards can be changed and whether there are distinctions between personal and professional values and ethical standards.

Even the marginal data summarized in this background paper, however, point to some of the barriers in efforts to initiate and institutionalize research-related ethics and values training in the sciences, social sciences, and engineering, whether on a voluntary basis or in response to mandates such as the NIH's 1989 requirement that all predoctoral and postdoctoral trainees supported by T32 or T34 National Service Research Award institutional grants "must receive instruction in the responsible conduct of research."

Our survey and interview data point to multiple reasons why faculty, graduate students, and deans report that their departments or institutions are not as active as they would like them to be in professionally related ethics and values training. Delineating these reasons at the level of specific departments or programs is a task that should be undertaken by those who have, would like to, or must incorporate such activities into their formal or informal curriculum. As we have learned in the project interviews, for example, some faculty believe that "ethics and values" is at best a peripheral or marginal aspect of research training, while others think that it is important but that they have no competence to instruct or advise students in this aspect of their work. Not surprisingly, many faculty also believe that their students can best acquire their professional values and ethical standards by "example" — through a largely latent transmission from advisor or mentor to advisee or mentee. This view, which our project and the literature on "mentoring" indicate is a widely and strongly held one, opens up another critical question: how much genuine advising and mentoring, in the "classical" senses of these terms, are taking place in graduate and professional education?

Prompted in part by the NIH training grant requirements, there has been a recent flowering of interest and initiatives under the rubric of "teaching research ethics." These are needed and laudable efforts, but further steps will have to be taken if they are to transcend occasional conferences, seminars, and workshops (which too often are attended by those already committed to and knowledgeable about the topics), and become an institutionalized part of the "everyday life" of research and graduate training.

The sources, clarity, and strength of the "messages" about the centrality of professional values and research ethics in graduate education will have an important bearing on how much long-term attention will be paid to these matters. There are several routes by which stronger, clearer messages can be conveyed. Within the university, these include faculty leadership in graduate programs, deans and other senior university officials and, very importantly, graduate students themselves. In keeping with the academy's tradition of departmental and faculty autonomy with regard to teaching, 43 percent of the deans in our survey stated that committing instructional time to ethical issues is a departmental decision. Another 50 percent indicated that there is an informal institutional expectation about such teaching, while 7 percent said that their university had a clearly stated or written expectation. Our faculty and student data on their departments' actual levels of activity, however, indicate that these expectations have not yet been realized to any substantial degree.

Outside the university, a stronger role could be played by professional associations. Many associations have been active in the area of professional and research ethics, but our survey and interview findings indicate that their messages are not reaching faculty and students very effectively, if at all. And, as witnessed by the NIH training grant requirements, research sponsors also can send powerful messages to universities and graduate programs about the importance of teaching and learning research ethics. As early efforts to implement the training grant requirements illustrate, however, there also is a pressing need to "train the trainers." Importing "ethics experts" for an occasional discussion of ethical issues in one's discipline may have short-term utility, but in the long run it perpetuates the marginalization of ethics and values in research and graduate education. Science, social science, and engineering faculty should not be expected to become experts in ethics and values studies. But increasing numbers of faculty may welcome — or be persuaded to appreciate — opportunities to learn more about the nature of professional values, about the ethical issues in their field and how they are being or could be dealt with, and about how to incorporate such materials into the various contexts in which they work and train their students.

Appendix: Project Overview

The Acadia Institute's graduate education project, which is being conducted with colleagues at the University of Minnesota, has been supported since 1987 by the NSF's Ethics and Value Studies Program and is co-sponsored by the AAAS Committee on Scientific Freedom and Responsibility, the Council of Graduate Schools, and Sigma Xi. The project's major research foci include: (1) the professional values held by faculty and doctoral students; (2) how students' professional values are shaped; (3) what students are learning and by what means about the types of value conflicts and ethical issues they may encounter in their future work settings and roles as researchers and teachers; and (4) the types of misconduct and other ethical problems that are occurring in graduate education programs and how they are being handled. We are examining these and other questions through a comparative study of faculty, doctoral students, and the structure, climate, and culture of their departments/ programs in four disciplines — chemistry, civil engineering, microbiology, and sociology — in major research universities.

The study's findings and recommendations are being generated by comparative and associational analyses of four sets of project data: (1) a 1988 national survey of graduate school deans concerning university policies and ethical issues in research and graduate education, which had a 66 percent usable response rate, (2) a 1990 national survey of 2,000 doctoral students (500 in each of the four disciplines) enrolled in 98 departments at major research universities, which had a 74 percent overall and 72 percent adjusted response rate, and (3) a 1991 survey of 2,000 graduate school faculty in the departments whose students were surveyed in 1990, which had a 62 percent overall and 59 percent adjusted response rate, and (4) a set of 78 in-depth interviews conducted during 1991-1992 with graduate students and faculty in the four disciplines at eight departments in three major research universities.

The rich body of quantitative and qualitative information that the study is providing should be useful to a number of audiences. For example, scholars working in fields such as ethics and value studies, higher education, and sociology will have a large body of data and analyses that will contribute significantly to intertwined areas such as professional ethics, professional or occupational socialization, and the climate and culture of academic disciplines and departments. Focusing more specifically on graduate education in the sciences, social sciences, and engineering, the study will provide graduate faculties, administrators, and students, as well as other relevant groups such as professional organizations, with information that will enhance their knowledge about the ethics and values context of that education. There is growing discussion and debate about the moral climate and responsibilities of universities; about the nature, extent, and handling of value conflicts and ethical problems within the academy; about the teaching, advising, and mentoring roles of faculty, especially in research-intensive universities; and about how these matters are affecting the content of the ethical "messages" being transmitted to the next generation of teachers and researchers. Our findings will inform this dialogue and, where specific problem areas are identified, provide bases for attempting to remedy them via modifications in curriculum, research training practices, faculty-student relationships, and graduate school policies and guidelines.

Teaching Ethics In Science

Vivian Weil, Illinois Institute of Technology

My topic is transmitting good research practices — the Why, the What, and the How. I will concentrate longest on the What. Why should we teach ethics in science studies? The first reason is that ethical choices are integral to scientific work. Would it be wrong to keep this new finding secret? From whom? For how long? The second is that there is now a mandate to teach ethics in graduate training in science. The National Institutes of Health and the Alcohol, Drug Abuse, and Mental Health Administration require an ethics component in training grants for graduate students. The third reason is that some recent empirical research indicates a serious lack of attention to ethics in graduate training in a range of fields.

Let me amplify a bit, taking the reasons in reverse order. Research reported in February, 1993 at the American Association for the Advancement of Science meetings indicates that there is a gap between students' expectations concerning appropriate research practices and what they observe around them in their labs, research groups, and departments. More importantly, the research also shows that faculty believe they are transmitting good research practices while their students report a lack of such training. Judith Swazey of the Acadia Institute reported these findings based on survey and interview research she directed under a grant from the National Science Foundation.

Before Swazey and her colleagues completed their research, NIH and ADAMHA recognized the need to ensure that ethics is part of graduate training. The federal agencies' requirements are an important source of some efforts already underway to teach good research practices and consider ethical issues in the practice of science. At a number of academic institutions, with little or no coordination between institutions, faculty have begun to try different methods for introducing ethics in science studies. So far only a few projects have been launched to share teaching materials and disseminate information about what works and what does not work.

The research and the government mandates generate an impetus to propagate the teaching in other institutions, to continue to innovate to produce new courses and new methods, to add to the materials for teaching, and to establish pathways for communicating and sharing the products of independent efforts. The aim of these efforts is to foster climates and practices that support responsible conduct.

I will turn now to issues to be included in teaching and methods to be recommended on the basis of our experience to date. We have learned from the teaching of practical and professional ethics, an enterprise that began with biomedical ethics in the early 70s. And the shorter period — barely five years — of teaching ethics in science studies yields insight and lore. Our sense of the topics to be covered derives from cases that have come to light and the ongoing literature on integrity and misconduct in science. The discussion of topics that follows should amplify my first point, that ethical issues are integral to doing science.

A distinctive feature of graduate training in science is that graduate students have a one-to-one relationship with a mentor or advisor (not always the same person), and students often, if not generally, have an exclusive association with a single laboratory. An important task is to figure out how to introduce explicit attention to ethics in these settings and how to broaden the exposure of graduate students to more scientists and labs. We can divide the issues to be dealt with into three categories: issues having to do with the everyday conduct of science, issues having to do with the structure of the institutions of science, and issues dealing with intersections of science and society. For an example of the first kind — the conduct of scientific work — take the matter of keeping laboratory notebooks. Among the reasons for keeping careful records of one's activities is the concern for accountability. Suppose the work is questioned later. For an instance of the second category — the structure of science — consider the peer review system and its associated roles and obligations. For an example of the third kind, think of the question of whether scientists have responsibilities to contribute to public debates about, say, global climate change.

With graduate students, the first category is the place to begin, and it will be the main focus of my remarks. It is important to make explicit the assumptions underlying everyday practice and shared expectations about how to carry out scientific research. The rationales for these assumptions and expectations should be examined. In the course of the examination, assumptions and expectations may be modified or even replaced. We have learned from a body of science studies by historians, philosophers, and social scientists that scientific practices are historically situated, shaped by local conditions, and therefore subject to change.

Before we proceed, some comment about misconduct is essential. The need to deal with cases and controversies in universities and in the government agencies that fund research led to efforts to define misconduct. The definitions promulgated by NIH and NSF list the core elements: fabrication, falsification, and plagiarism. A fourth category encompasses "other serious deviation from accepted practices in proposing, carrying out or reporting research." This fourth category has understandably become highly controversial and is apparently on its way out of the NIH definition. It was included because both the Public Health Service and NSF found they needed it to cover some cases they had to investigate.

Teaching has to include some attention to the elements of misconduct at some point. One might think that cheating is wrong and there is no more to be said. But determining in some complex circumstances that falsification or fabrication has occurred turns out to be less straightforward than one might expect. Plagiarism needs careful attention. It is not easy to explain what is wrong with plagiarism when it does not involve copyright infringement. There are other questions. How do we distinguish plagiarism from poor citation practices, "half-way plagiarism?" Is it plagiarism to use the refinement of a technique developed by another graduate student in the lab working with the same mentor? To answer, it will help to look at a range of examples to see what are the values served by the condemnation of plagiarism and what are the identifying marks of plagiarism.

Scientists like to emphasize that misconduct should be distinguished from error. Error, they say, is part of science, and they know how to counter it. However, error too needs to be probed. Some errors are more serious than others, and some are culpable. Ethical codes and guidelines are one resource for dealing with this issue, and others. In November of 1991, the American Physical Society published *Ethical Guidelines* which state that scientists may be responsible for failing to acknowledge and correct error.

Sloppiness and carelessness are also important to consider. Sloppiness can amount to recklessness or incompetence or breed intentional wrong doing. Sloppiness is of concern because it appears to be more common than other lapses, and the harm from sloppiness may not be sufficiently appreciated.

The fourth element of misconduct, serious deviation from accepted practice, leads us away from concern with misconduct — important as it is — and leads us instead to examine research practices and standards or rules of practice. Many people count on informal standards to assure good practice and provide a barrier to misconduct. Standards vary across research communities, but, as the research indicates, in a specific community, they may be difficult to discern.

Specific practices have been identified for examination in teaching and research. The list includes authorial practices, mentorship practices, keeping lab notebooks, handling data, writing reports, and data sharing. Some of these practices are tied more closely to structures of science than others. Data sharing, for example, has been the subject of funding agency rules. But all these practices should be considered locally in research communities, groups, and laboratories. At my own institution, in a faculty research ethics sack lunch group that has been meeting once a month for a year, many have commented on their discovery of the need to make expectations explicit.

After I comment briefly on specific practices to be covered in teaching, I will survey some methods and conclude with a "case" useful for teaching. Authorial practices, especially practices of allocating credit, raise issues about misrepresentation and about personal responsibility. Honorary authorship — the practice of automatically including the name of the director of the lab, for example — is subject to question even when the convention is well established and defended on the ground that it provides useful information. That is because responsibility hinges on a clear link between authorship and writing the paper or contributing substantially to the creation of the paper or its contents.

The recent research I alluded to at the outset and our heightened awareness of the vulnerable position of graduate students, postdocs, and clinical research trainees make clear the importance of giving attention to mentorship issues. These issues include the proper ratio of trainees to mentor, training structure, choosing and allocating research problems, and conflict of interest. This is a short list. As we continue to collect cases, we find additional aspects of the relationship to consider. For example, some vignettes contributed by graduate students show that submitting abstracts under deadlines is a situation that can generate problems in the mentor/student relationship. The maintenance of laboratory notebooks came into prominence in the Baltimore case. Questions concern determination of who records data, in what form, under whose control notebooks are kept, and who retains ownership. Again, we find a variety of practices across research communities.

Problems concerning the treatment of data include the handling of outliers, determining what data may be excluded, and deciding when an experiment is complete. Justifications for excluding data need to be thoroughly discussed within research communities. For an indication of some subtleties, notice that what counts as data is a matter of convention. Anthropologists, for example, do not count field notes as data. Data are findings that a researcher stands behind or gives authority to. Someone takes responsibility. Data is thus a value-laden notion. The writing of reports, though closely allied, deserves separate consideration. Allocating emphasis and devising graphs and charts are among the topics to include.

Issues about openness and secrecy are important to consider in light of scientists' traditional commitments to openness and the barriers to openness that arise in scenes of intense competition. It becomes necessary for research groups to formulate policies that encourage openness, taking into account needs for secrecy that can be ethically justified. In this connection, issues surrounding intellectual property could receive attention.

Turning to the structures of science, we note that peer review, journal editing, and the grants process are systems that have received some scrutiny. It is useful to consider the many purposes served by peer review and then to give attention to responsibilities of participants in those enterprises. The purposes include, among others, improving research, establishing research priorities, focusing skepticism, allocating scarce resources, supporting professional authority and autonomy, and contributing to a sense of collective enterprise.[1] Ethical dimensions of duties associated with the roles of reviewer, journal editor, funding agency program director, and others need to be discussed.

Now I turn to the How. Cases or scenarios of everyday, problematic situations are widely regarded as critically important in both research and teaching. They are inevitably the point of departure in practical and professional ethics teaching. Suitably fleshed out with detail, they can serve in informal discussion or in a course on research ethics to bring out a problematic situation calling for a response. It is important when devising a response, say, to uncertainty or conflict about the status of a lab notebook, to consider whether this response could or should be made a policy, perhaps even included in guidelines. In this way local practices can be made explicit and scrutinized with an eye to revision. In our experience at my own institution, the exercise of case analysis brings out subtleties of the circumstances and opportunities for careful handling of situations that can forestall conflict and help to build trust. This exercise can go on in a lab, department, research group, or with a group of faculty, postdocs, or students from a range of fields. It is useful to devise seminars or other opportunities for people in science to confront the variety of practices across fields.

I will conclude with an example. The seven step procedure for case analysis (Figure 1) is useful to contain discussion and reach closure. It is a kind of checklist, not a mechanical decision procedure.

Graduate student named Lee worked with Professor Heisman in a highly regarded lab on project Plum. By the end of that year, Lee had not only become proficient at many of the more routine tasks of the project but had made a small but notable refinement to the approach in the segment assigned to Lee. At the end of the first year Professor Heisman went on leave for a semester and Lee started working with Professor Kaltman in the same laboratory but on a very different project. Professor Heisman returned for the spring semester and took up the Plum project, among others. The following fall, the beginning of Lee's third year, Lee learned from another student, Leslie, who was working on Plum that Professor Heisman was publishing a paper on some aspects of Plum with Leslie (only), a paper which contained Lee's refinement.

What, if anything can and should Lee do? What should Lee do as a first step in light of any ambiguities in the situation? What should Lee do if those initial efforts fail to achieve the desired results?

Acknowledgement:

This essay is based on and draws from an earlier essay entitled "Ethics in Scientific Research and Graduate Education" that appeared in *Scientific Responsibility and Public Control*, Centre for Research Ethics, Göteborg, Sweden (1993).

Notes:

1. This list is drawn from a set of ten purposes set forth by Professor Ed Hackett of Rensselaer Polytechnic Institute in a talk at the 4S/EASST meetings in Göteborg, Sweden in August, 1992.

2. This example is adapted from a case offered in an e-mail discussion (November, 1992) by Caroline Whitbeck of MIT.

A Seven-Step Procedure for Moral Decision Making

1. Recognize and Define the Moral Problem.

2. What Are the Facts?

3. Identify Affected Parties.

4. Formulate Alternatives and Continue to Check Facts.

5. Assess the Alternatives

    a. The ethical implications of alternatives.

    b. What are the practical constraints?

6. Construct Desired Options and Persuade or Negotiate With Others to Implement Options.

7. What Action Should Be Taken?

    (Go back to Steps 1 and 2 to see whether this solves the moral problem and check whether any facts were missed.)

Figure 1

Introduction to the Societal Responsibilities of Science

Chauncey Starr

Preface

The role of science as an influence on societal actions has increased markedly in the past century as our society has become more industrialized and more shaped by technical systems for supplying goods and services. In earlier periods when science focused on basic cause/effect relationships for a few natural phenomena, it could only provide a minor factual input to societal decision-making. Empirical relationships and their applications to the practical arts (engineering, medicine, military) usually preceded scientific understanding. Everyone knew water ran down-hill long before a gravitational force was understood. Practical inventions were mostly the result of cut-and-try empiricism, sometimes termed the "Edisonian" method. So the scientific community was relatively distant from the societal consequences of technology's early growth. During this century, the direct influence of science on societal actions has gradually grown, and today it carries an increasing share of the responsibility for social decisions based on science and technology.

What is Science?

Prior to this century, science was predominantly a communal activity of scholars in which theory and experiment about natural processes interacted to correct errors in both hypotheses and experiments, and finally produced a few verifiable relationships. In this century, science joined with the Edisonian method to provide the practitioners of engineering the design bases for technical systems.

Unfortunately, an unintended byproduct of the large growth of basic science research, and thus an accelerated rate of new scientific hypotheses and experimental findings, is that today the slow and costly process for verification of new ideas is overwhelmed. Verification involves replication, search for confounding variables, and testing by prediction or trial. Historically it has taken many years, or decades, to confirm or disprove the validity of new findings. While much of recent science remains within the scientific community for internal debate (e.g., astrophysics), many new findings are immediately relevant to many public concerns, such as living quality, economics, health and safety. The rise of environmental and consumer activism on such matters since the early 1970's has placed greater emphasis on the early interpretation of scientific findings. Thus the uncertain validity of new but untested findings, and of the societal decision-making response to such uncertainty, involves the relevant scientific community. This is a new role for scientists and no process exists for fulfilling it wisely.

What is not Science?

"Things are seldom what they seem

Skim milk masquerades as cream"   (G&S 1879)

A layman's concept of the scientist as a white-coated oracle of truth seems obviously naive to any practicing scientist. Unfortunately, it is an image displayed by the media, romanticized in literature, and quietly encouraged by scientific organizations for professional prestige enhancement. As a result, every survey places "scientists" high in the public credibility list of authoritative sources. However, when a scientist uses this authority image to ennoble his role as a citizen when supporting a public policy, the potential for societal irresponsibility exists.

Science does not embody social activism. While many scientists as members of the public enthusiastically urge various social policies, and may use their scientific knowledge as support, their contribution to the public debate is that of informed laymen — but not as spokesmen for the scientific community. Unfortunately, special interests, notably the environmental movement, have involved many scientists in their causes, and have used them to convince the media and politicians that scientific certainties support their programs when, in fact, the science is uncertain. It is this "white-coat" pretense which is irresponsible. A skeptical media and public should carefully weigh the scientific claims of these special interests — a difficult task for laymen or for the communication media.

In the early days of scholarly science, the field was chosen as a personal career for its intellectual rewards, like art. Today, science is an institutionalized industry. However, science is not an entitlement program for scientists, demanding society's support. The support of science should be deserved by its contributions to our national needs, and balanced against other social goals. This is especially the case when the scientist's personal welfare is clearly involved in the promotion of projects. For example, it appears disingenuous for NASA space scientists to suggest that a multi-generation program to populate the Moon or to land on Mars should have a high national priority, at the expense of other social investments. Such a self-seeking ploy to maintain the manned-space program should have much discussion within the scientific community and the public forum before political acceptance as national policy. If other than scientific values are involved, their merit should be publicly evaluated. There are many such debatable programs underway in the U.S., and the scientific community appears to have avoided its responsibility for providing a balanced perspective to the public on these. Among scientists it is commonly understood that criticism of other scientists invites retribution in peer reviews and grants, and with intuitive political instinct they usually avoid such criticism. Does this mean that the scientific community has no voice or avoids the responsibility to use it?

What is the scientific information process?

Science probes the frontiers of knowledge about nature, and its findings and speculations are developed as new ideas. Their disclosure to the scientific community requires merchandising these new ideas by publications, meetings, seminars, and peer reviews by various granting agencies. Then follows a slow process of collegial verification. The outcome provides three levels of validity; (1) verified or negated by replication and test; (2) plausible but incomplete and not fully tested; and (3) possible but speculative, uncertain, and not testable. Basic physics and chemistry are examples of the first. The second describes most large-system models that are constructed by the synthesis of many sub-systems; examples are weather models and medical models of disease processes. The third applies to most predictive models intended to forecast the behavior of complex natural systems; such as earthquakes, climate futures, epidemics, mineral resources. The chief scientific function of such models is to suggest promising directions for research that might usefully add to our body of knowledge. Their common characteristic was best voiced by Buckminster Fuller (1966) "Synergy means behavior of whole systems unpredicted by the behavior of their parts."

So science manufactures and merchandises ideas. As long as these remain within the scientific community, the internal processes will establish their appropriate level of validity, and the direction for further study. Only a fraction of new findings eventually meet the test for validity. But if categories 2 and 3 are placed in the public domain and are seriously considered as a basis for government strategies, the social responsibility of science to place these in proper perspective exist. While the public is appropriately skeptical about merchandising claims for goods, and the media more so, both appear to be gullible when a scientist speaks. Premature science communication can do harm as well as good.

It should be recalled that government strategies change exceedingly slowly, and whatever course is taken, a parasitic constituency develops whose survival opposes change. A classic example is the infamous Delaney Amendment of 1958 requiring "zero risk" for food additives — a populist goal with strong emotional content but without any scientific basis and impossible to prove. The amendment arose from the simplistic hypothesis that by complete removal of man-made carcinogens, one-by-one, cancer risk would eventually disappear —obviously a popularly appealing notion. For 35 years every government agency affected by this amendment has sought a way around its unrealistic constraint, and yet Congress has not had the wisdom or courage to remove it. The amendment has been used in legalistic procedures to force extreme agency actions. Sugar substitute regulation is a well-known example. The risk analysis community has regularly expressed its dismay at this situation, but to no avail.

Are all scientists biased?

It should be recognized that all scientists have varying levels of bias when dealing with non-verified hypothesis, models, and predictions. The normal professional act of merchandising of such concepts by publication and presentations makes the scientist a salesman for them. Bias results from emphasizing the positive aspects of a concept relative to the negative. The degree of bias, whether conscious or unconscious, depends heavily on the goals, rewards and penalties that shape the framework of the expert's views. The academic researcher seeks the accolades of his professional colleagues, and depends on such recognition for promotions, tenure, professional awards, and most importantly grants from government agencies, foundations, and industry. The politics of science reveals the extent of manipulations by individual scientists to achieve these rewards. For most scientists non-monetary recognition is as strong an inducement as research support or industrial salary. The recent flurry of investigations into the ethics of scientists discloses the great value placed on the non-monetary goals. With such inducements a scientist would need to be a saint to avoid some degree of bias when merchandising his concepts.

Expertise in scientific specialties resides in those who work in depth in their fields, whether in academia or industry. Such work almost always requires substantial financial support, either from government agencies and foundations or from industrial organizations. All these funding groups have goals that they assume will be furthered by the scientists they support, even though all avow they seek only the truth. It is generally recognized in academic circles that government agencies are not likely to support those whose opinions might weaken the agencies' budget submissions to their government. Similarly, industry is very uncomfortable with research findings that questions its public positions. These pressures further stimulate scientist's bias, often unwittingly.

In view of such bias, how do scientists arrive at an objective consensus on the validity of category 2 and 3 topics? The individual views and opinions that arise from each scientist's working relations are best disclosed and balanced by assuring a mix of experts. It must be recognized that every scientist, whether in academia or industry, places maximum value on professional credibility. For this reason, in a mixed expert group a consensus on fact finding is usually arrived at without stress. Differences usually occur in the area of uncertainties and their policy implications, and it is here that the virtue of a mix of scientific backgrounds becomes evident in disclosing the range of issues for consideration by policy makers. However, the now common practice in government appointed committees of combining a mix of scientific experts with a mix of policy advocates abrogates the objective of a scientific consensus on the degree of validity and significance of scientific findings. Policy advocates have sought such participation as a means of biasing the outcomes for their causes, and have successfully persuaded many agencies that this is needed to balance the presumed bias of expert scientists. This clearly politicizes many agency studies and reports.

What is socially responsible science?

It is evident that science makes its most responsible contribution when closely adhering to verifiable knowledge about nature (category 1). This has historically been the basis for most advances in industrial countries, and continues to be supported by financial feedback from the national economy. It involves the bulk of all scientific activity. The issue of social responsibility in the application of category 1 science arises only in the rare circumstances when some entity, government or industry, incompetently violates a proven relationship and thus creates a public risk. Incompetence is a management responsibility.

The social responsibility of science is most evident in the participation of the synthesizers and model makers in public decision-making. The plausible but incomplete and not fully tested system concepts (category 2) are a common part of many national activities requiring near-time operating decisions. For example, space launch decisions of NASA, or military decisions of the DOD, or regulations of EPA, involve somewhat uncertain outcomes and risks, and the actions undertaken are a judgmental consensus of scientists, agency managers, politicians, and sometimes Congress. If failures occur as a result of a scientific error, the responsibility is quickly focused. However most failures are due to scientific uncertainty. The scientists involved may not have adequately disclosed their areas of ignorance, and have underrated the risks of failure in their enthusiasm for the mission. This does not reflect on their sincerity or integrity, but it does place in doubt their scientific credibility. Such situations are common in all government agencies, industry, and special interest groups, as enthusiasm for a mission is a desirable administrative criterion.

The most serious issue of social responsibility arises from public decision-making based on the speculative, uncertain and untestable model projections of future outcomes (category 3). History has shown these to have been notoriously unreliable, yet they are sometimes so dramatic as to excite public fears and social mitigation strategies to control the future. Many scientists enjoy and profit from the resulting public excitement. In fact, an established "futures" constituency now exists with a mix of science consultants, special interest groups, academics, and lobbyists symbiotically interacting with politicians and government agencies. Under the virtuous mantle of saviors of humanity and the earth, each group pursues its own agenda for shaping the future. Nevertheless, there is usually within such future concepts elements of validity which might be useful guides for near-term decisions. Responsible science would focus on extracting these, rather than promoting the flamboyant dramatics so loved by the media, and so frightening to the public.

Examples of category 3 phenomena abound. Predictions of Malthusian catastrophes have been the most common, since a simple-minded exponential (compound interest) growth of anything will lead to quantitative enormities. In the early 1900's, agricultural scientists confidently predicted a U.S. shortage of arable land would result in imminent mass starvation unless immigration was curtailed. More recently we have had the scare of a global "nuclear winter" — now discredited as more research developed. In the 1970's the fear of a future scarcity of natural gas resulted in the Fuel Use Act that prohibited its use by utilities for electricity generation. It was recently repealed as it became evident that we have a natural gas surplus, and the utilities are now being encourage to shift to gas as the most benign fossil fuel.

The more recent movement to balance protection of the environment with economic growth, laudable as it may be, has provided a platform for many dire predictions with very weak scientific underpinning. Today's most fearsome long-range threat is the prediction of serious global climate change from greenhouse gases, with carbon dioxide considered the most important. This possibility was mentioned early in this century, but in the absence of any obvious climate change, it remained quiescent until the last few decades when the environmental consequences of increased energy use were brought to the fore. Subsequent climatologic research has been disclosing pieces of the puzzle on a monthly basis. For example, the early and frightening prediction that a large sea-level rise would inundate huge coastal areas have now receded in plausibility as the latest studies suggest that an ocean temperature rise would probably increase precipitation on the polar ice-caps and thus lower the mean sea level. The climate effects of increased cloud cover and aerosols remain uncertain, but most recent results suggest they may reduce climate change. It is clear that the dire predictions of drastic climate changes are very uncertain and scientifically premature as a societal guide. Does this mean that no attention should be given them? Hardly. Within the science community they should be seriously evaluated to provide useful insights for further climate study, so as to gradually reduce the uncertainty. And the political urgings for more efficient use of nature's resources and less emission of waste gases certainly have merit for many reasons. This is but one example of the interaction of category 3 science with the actions of social decision-makers and strategists. Industrial societies are replete with such category 2 and 3 interplay with political actions.

With the goal of transparency, responsible science calls for full disclosure to decision-makers and stakeholders of the state of knowledge and the risks associated with areas of uncertainty. Most scientists are aware of these issues, but frequently succumb to the pressures mentioned above to bias their judgment. The guiding principle for the responsible scientist is "a half truth is a whole lie." It is rare for any scientist to knowingly lie — it could be career threatening. However, a half truth implies suppression of opposing information which may be important in the decision process.

Can social responsibility be achieved?

The overriding issue of social responsibility for science is how to provide the public decision-processes with a balanced perspective of the validity of the scientific input to that process. This must be an undertaking of the scientific community, as individual scientists have all the frailty and as well as the nobility of the common man. The scientific community needs to establish means for undertaking this task as a step to develop a socially responsible role. The recent methodology for the comparative risk analysis of alternatives provides a basis for such a role, but is rarely practiced. Competently performed it is so revealing that only a few single-issue programs would remain unscathed. For some time the Office of Management and Budget has urged agencies to undertake this process, with little success. What promoter wants the "naked truth" disclosed?

Several decades ago a "science court" was suggested for this purpose, but not supported, probably because legal procedures are anathema to scientists — scientific truth is not determined by a vote of judges or juries. National Academy committees are a partial step, but they now function only at the behest of government agencies, and depend on such agencies for financial support. Nevertheless, they function reasonably well although slowly. They may provide an initial step for the scientific community, supported by the participation of the purely scientific societies. The community needs to develop a credible voice to serve the public. In the meanwhile, we must recognize that those intent on manipulating public opinion profit from the existence of a malleable and muted scientific community.